The Changing Role of Science Centres and Museums in Developing Countries

The Changing Role of Science Centres and Museums in Developing Countries

Edited by

Dr. Narender K. Sehgal

2005
DAYA PUBLISHING HOUSE
Delhi - 110 035

ISBN 81-7035-379-3

Centre for Science and Technology of the Non-Aligned and Other Developing Countries (NAM S&T Centre)
Core-6A, 2nd Floor, India Habitat Centre, Lodhi Road,
New Delhi-110 003 (India)
Phone: +91-11-24644974, 24645134, Fax: +91-11-24644973
E-mail: namstct@vsnl.com
Website: www.namstct.org

Partial Financial Support by : **Vigyan Prasar**
C-24, Quatab Institutional Area, New Delhi-110 016
Phone: +91-11-26864157, 26967532, Fax: +91-11-26965986
E-mail: vigyan@hub.nic.in
Website: www.vigyanprasar.com

Published by : **Daya Publishing House**
1123/74, Deva Ram Park, Tri Nagar, Delhi - 110 035
Phone: 27383999, Fax: (011) 23244987
e-mail: dayabooks@vsnl.com
website: www.dayabooks.com
Showroom : 4762-63/23, Ansari Road, Darya Ganj
New Delhi - 110 002
Phone: 23245578, 23244987

Laser Typesetting : **Classic Computer Services**
Delhi - 110 035

Printed at : **Chawla Offset Printers**
Delhi - 110 052

Printed in India

Foreword

The social and economic development of a country is closely linked with the understanding and application of science and technology by its people. It is hence imperative that we comprehend and understand the impact of scientific and technological advances that have shaped our lives in the past few decades. In this respect, science centres and museums play a vital role in communicating basic scientific information to the people and help them make informed choices. Further, science centres and museums act as an interface between the general public and the scientific community. Indeed, science centres and museums are powerful tools for social change and for developing a scientific outlook among the people.

I am happy to note that there has been a growing consciousness among the developing countries about the role played by science centres and museums in educating the people and in the improvement of the quality of life. Indeed, NAM S&T Centre has long recognized the need and importance of science centres and museums, and their relevance to developing countries. Efforts of NAM S&T Centre, therefore, are noteworthy in bringing together science popularization experts and museologists from the developing countries on a common platform to share their knowledge and experiences. I am especially happy that the papers presented and experiences shared during the Second International Workshop on the "Changing Role of Science Centres and Museums in the Developing Countries" organized by NAM S&T Centre during 20-22 October, 2004 at Hanoi, Vietnam, are being brought out in the published form. This volume, I am sure, would prove to be of immense utility for initiating and sustaining a people's science movement in the developing countries.

I congratulate Dr. A.P. Kulshrestha, Director, NAM S&T Centre for his sustained efforts in promoting mutually beneficial collaboration and cooperation among scientists, organizations and institutions from Non-aligned and other Developing Countries and transfer of technology among them. I am also happy to note that the present volume is being published jointly by Vigyan Prasar, New Delhi, and NAM S&T Centre.

V.S. Ramamurthy
Secretary
Department of Science & Technology
Government of India
New Delhi – 110 016

Preface

Nearly all the presentations made at the Hanoi Workshop, and included in this Volume, are from individuals who are either working, or have at one time or another worked, in a science centre or a science museum. Not surprisingly, therefore, the content of almost all the presentations revolves around programmes, activities, galleries, displays, exhibits, methods, means, strategies...to attract more and more visitors day in and day out. As per the focal theme of the Conference, i.e. "The Changing Role of Science Centres and Science Museums", some of the presentations also discuss the shift of focus from "do-not-touch" exhibits and displays to more and more "hands-on" and "do-it-yourself" items, as also those which the visitors could assemble or dismantle. Most of the authors are agreed on the proposition that science centres and museums are places for continuing informal education and an essential role for them is to complement and supplement school education.

While editing this book, I also attended the 4th World Congress of Science Centres and Science Museums in Rio de Janeiro, Brazil (11-14 April 2005). One of the issues that was being emphasized there was that the Science Centres/Museums around the world, as of now, required their visitors to have a minimum level of grounding in science in order to benefit from what they had on offer. But in most developing countries where literacy was not yet high enough, or there were not enough schools for all children, there was need to have non-formal "pre-science centres" which could work with out-of-school children, or school drop-outs, to fill the gap that the present science centres/museums were not equipped to do. Is this another way of saying that the concept of a science centre, or a science museum, lifted straight out of the developed world, may not be fully suitable, fully appropriate, or "full value for money", for developing countries? Something to think about, seriously, for the next conference!

While we have an NCSM (the National Council of Science Museums) in India, we also have an NCSTC (the National Council for Science and Technology Communication). While the former works on and promotes the use of science centres and science museums as tools for informal science education and science popularisation, the latter works with and on all other methods and means of "popularising science & technology and promoting stimulation of scientific and technological temperament among the people". NCSTC (set up in 1982) has over the years built up a comprehensive programme of science and technology popularisation which has notched up pioneering achievements. While working parallely, NCSTC and NCSM have also been collaborating with each other on very large scale countrywide projects, reaching crores of people. My bringing out the case of NCSTC here has one sole purpose: To stress that while science centres and science museums do have a very important role to play in popularising science, there is also another, less expensive line of approach which all developing countries can greatly benefit from–and, incidentally, it happens to be more appropriate and more suited to their needs and ground realities.

Before closing, I would like to mention that it has not been easy to edit this Volume, because of the wide variation in writing styles and the idiom employed by presenters from different countries. A lot of time and effort have been invested in going through each presentation and in making the minor alterations that were made. I do hope that, in the process, I have not earned any presenter's ire. My apologies, if I have.

Narender K. Sehgal

Contents

Introduction

Science centres and museums play a key role in discovering, exploring and testing ideas about the natural world and helping to forge a more confident relationship between science and society. Such assertive links are imperative because though science and technology can offer many benefits to the society and to the economy, these benefits can only be realized with the support of the public. We are on the brink of exciting developments in science that will affect everyone's life. However, along with the benefits come potential risks that cause understandable concern. People rightly expect proper safeguards against risks, and full and open discussion about the values, benefits and risks of scientific developments. Science centres and museums are important to taking forward this subject and engaging the public with science and the issues that it raises for society.

The role of museums in society has expanded. Not only are they devoted to the procurement, care, study and display of objects of lasting interest or value, but also are safe havens where individuals and families can connect with others. They are the information centers for children and adults offering resources that give context for understanding our world and special environments where people may retreat into work, beauty, literature, entertainment, projects or hobbies. Science centers and museums offer a wonderful opportunity for raising public awareness of current research, providing a neutral venue for debate and dialogue, and stimulating an interest in science among young people. These can also help in encouraging more people from ethnic minorities into science. This challenge is an extremely important one if we are to realize the benefits of science. It will enable all of us to exchange ideas and forge new partnerships, and make a critical difference in this important dimension of our national lives. In order to fully comprehend the unprecedented growth in various fields of science and technology and their impact on the socio-economic progress, particularly in an emerging economy, the role of science centers and museums is vital because of visual impression aids in the process of quickly grasping various facets of the new knowledge.

The Centre for Science and Technology of the Non-Aligned and Other Developing Countries (NAM S&T Centre) had organized an international workshop on 'The Role of Science Centres in Development Strategies' in Kolkata, India, during November 30–December 02, 2002, at the conclusion of which a Kolkata Declaration was adopted that emphasized on holding a general workshop on science centers and museums once every alternate year with the involvement of experts from the developing countries of the entire globe, and a regional workshop during the intervening period. In pursuance of the Kolkata Declaration, a Latin American and Caribbean regional international Forum and workshop for Culture of Science, Technology and Innovation in Society and the 1st Knowledge Fest was held in Bogotá, Colombia, during 24th-26th March 2004 with the enthusiastic efforts of Dr. Elizabeth Hoyos, Director, Maloka Science Centre of Bogotá.

During October 20-22, 2004 an international workshop on 'The Changing Role of Science Centres and Museums' was organized by the NAM S&T Centre jointly with the Ministry of Science and Technology of Vietnam (Centre for Regional Research and Development, CRD) in Hanoi, Vietnam, to review the status of science centers and museums and related establishments in the world, particularly in the developing countries and to exchange ideas and experiences pertaining to their present and future developments and cooperation. The event was focused at encouraging such institutions in the developing world to prepare for greater demands on their services. Representatives from eleven countries, namely, Algeria, Egypt, India, Indonesia, Malaysia, Mauritius, Nepal, South Africa, Sri Lanka, Vietnam and Zambia participated in the workshop.

The Hanoi workshop was spread over six technical sessions besides the inaugural session. From planning programmes to securing funds, to securing facilities and reviewing, these sessions addressed a variety of issues pertinent to science centres and museums today. H.E. Dr. Bui Manh Hai, Vice-Minister, MOST, Vietnam, Prof. Arun P. Kulshreshtha, Director, NAM S&T Centre and Mr. Phan Huy Chi, Director, CRD, MOST, Vietnam, jointly inaugurated the workshop. The theme papers in the technical sessions were presented by Mr. Phan Buy Chi, Director, CRD (MOST), Vietnam; Mr. Mesbahul Alam, Scientific and Technological Advisor, Ministry of Science and Information & Communication Technology, Bangladesh; Prof. Dr. Salwa Abdalla El-Gharib, Faculty of Applied Arts, Helwan University, Egypt; Dr. V. B. Kamble, Scientist 'G' and Director, 'Vigyan Prasar', Department of Science and Technology, Government of India; Dr. J. Mishra, Programme Associate, NAM S&T Centre, New Delhi, India; Dr. Tran Ngoc Ca, Vice-Director, National Institute of S&T Policy and Strategy Studies, MOST, Vietnam; Ir. Herudi Kartowisastro, Director, Indonesian Science Centre, Jakarta, Indonesia; Mr. Viswamitra Oree, Curator, Rajiv Gandhi Science Centre, Mauritius; Prof. Dayananda Bajracharya, Vive-Chancellor, RONAST, Nepal; Prof. Dr. Le Ngoc Ly, Department of Oceanography Naval Postgraduate School and Advisor, CRD, Vietnam; Mr. Ratan Kumar Chakrabarti, Education Officer 'C', Birla

Institute of Industrial Museums (NCSM), India; Mr. Samir Kumar Ray, Director, CTRL (NCSM), Kolkata, India; Mr. Jayanath Appudurai, Director of National Science Centre, Ministry of Science, Technology and Innovation, Malaysia; Mr. Nguyen Van Huy, Director, Ethnic Museum, Vietnam; Mr. Arumahennadige Piyasiri, Mechanical Engineer, National Engineering Research and Development, Sri Lanka; Mr. K. M. Lebs Mphahlele, Manager of Science and Youth, Department of Science and Technology, South Africa; Dr. Nguyen Ba Thi, National Institute for Education Strategy and Curriculum, Vietnam; Prof. Dr. Le Ngoc Ly, Department of Oceanography, Naval Postgraduate School and Advisor, CRD, Vietnam; Mr. Festers Hangandu Mungo, Principal Education Officer, National Science Centre, Zambia; and Mr. Luu Ngoc Anh, Officer, Educational and Scientific Program Department, Vietnam Television. The workshop ended with the adoption of Hanoi Resolution 2004 signed by 14 country representatives respectively from Algeria, Bangladesh, Egypt, India, Indonesia, Malaysia, Mauritius, Nepal, Pakistan, Sri Lanka, South Africa, Sudan, Vietnam, Zambia and one representative from NCSM and BITM, India which was endorsed by the distinguished participants from various agencies and organizations from the host country, Vietnam.

The present publication comprises a compilation of the presentations made during the Hanoi workshop. The NAM S&T Centre would like to thank Mr. Thach Can, Director General, Ministry of Science and Technology (MOST) of Vietnam, Dr. Thai Van Tan, Director, UNESCO & Asia-Pacific Desk, Department of International Relations in MOST, Mr. Phan Huy Chi, Director, Centre for Regional Research and Development (CRD), Hanoi, Vietnam, and his colleague, Ms. Ngo Thi Loan, for shouldering the responsibility and sharing the finances to organise this important event jointly with the Centre.

The Centre expresses its sincere gratitude to Dr. Vinay B. Kamble, Director, 'Vigyan Prasar' and Scientist 'G' in the Department of Science and Technology (DST), Government of India, for arranging a financial contribution for bringing out this publication. The Centre also places on record the contribution made by Dr. Narender K. Sehgal, former Advisor DST and presently Chairman of the National Organising Committee, Year of Scientific Awareness 2004, of India, for his untiring efforts in editing this publication. The dynamic involvement of the entire team of the NAM S&T Centre in bringing out this publication is also gratefully acknowledged.

I hope that the material presented herein would serve as a permanent reference material for the professionals working in the area of science centers and museums and will help developing countries in working out appropriate plans and strategies for inculcating scientific values and temper among the masses.

(Arun P. Kulshreshtha)

Director, NAM S&T Centre

1

Status of Science Centres and Museums of Bangladesh: An Overview

Mesbahul Alam

Scientific and Technological Adviser (STA),
Ministry of Science and Information & Communication Technology,
Government of the People's Republic of Bangladesh,
Building # 6, 9th Floor, Bangladesh Secretariat, Dhaka-1000, Bangladesh
Telephone: 88 02 7167274, Fax: 88 02 7169606
E-mail: most@bangla.net, mesbah002@yahoo.com

ABSTRACT

Role of science centres and museums in Bangladesh is not so strong in disseminating science and technology among the masses. National Museum of Science and Technology (NMST) and newly established Bhasani Novotheatre (Space Theatre), Dhaka, under the Ministry of Science and Information & Communication Technology are the only science centres and museums in Bangladesh. Present activities of these centres are very limited and there is enough scope to enhance and develop the services presently being delivered to the people. Knowledge learned from the workshop will help the Ministry of Science and Information & Communication Technology to evaluate and reformulate the present policies and strategies of the science centres and museums to enhance their services. This paper provides a brief overview of activities and services of the science centres and museums in Bangladesh.

INTRODUCTION

Science centers and museums are unique resources for informal learning. They are the places to discover, explore and test ideas about the natural world engaging the public with science & technology. Science centres are places where people of all ages can learn at their own pace, engage their curiosity, use their senses to ask and answer questions, and explain to others what they have learned. Science centres and museums play an increasingly important role in bringing science and technology to the public.

Science centres connect people with science, give science a presence in the community and offer people of all ages and backgrounds the opportunity to ask questions, discuss and explore. In science centres you will encounter hands on exhibits like giant levers, wave tanks and walk in kaleidoscopes. While you are there, you may go to the demonstration, watch a sky show or big screen film, participate in a workshop or even take part in a debate about a current issue like bioethics. In the process, you will experience the pleasure of life long learning, whether with family, friends or on your own.

In recent years we have witnessed unprecedented growth in various fields of science and technology. In order to fully comprehend these developments and their impact on the socio-economic progress, particularly in an emerging economy, the science centres and science museums play a vital role because visual impression aids in the process of quickly grasping various facets of the new knowledge.

COUNTRY PROFILE: BANGLADESH

Bangladesh is a small country with 1,44,000 square kilometres of land located in the southeastern part of South Asia. India and Myanmar are neighboring countries. With this small territory, the total population of Bangladesh is 140 million. The density of population is nearly 2200 per sq mile–probably the highest in the world except city-states like Singapore. Most of the people live in the villages–rural: 85% urban: 15%. Population growth rate is also high in comparison with many developing countries. It's a deltaic riverine country, coupled with high rainfall (average 100cm) and hot-humid climate, which make it very fertile and cultivation friendly. Rice is the major food crop, which is cultivated round the year. Per capita income at present is under US $375(2002).

CURRENT STATUS OF SCIENCE CENTRES AND MUSEUMS IN BANGLADESH

In Bangladesh only two science centres and museums are currently providing services for the 140 million citizens under the Ministry of Science and Information & Communication Technology. Aims, objectives and services they are providing and future development plans are described in what follows.

THE NATIONAL MUSEUM OF SCIENCE & TECHNOLOGY (NMST)

The National Museum of Science & Technology (NMST) is an informal Science & Technology Education Centre. For dissemination of scientific and technological knowledge the NMST is playing the role of a Science Centre in Bangladesh. Its main function is to popularize science & technology among the common masses especially among the young generation for creating interest about science and information & communication technology. For performing the said function the NMST arranges guided tours in it's different galleries for explaining scientific information and theories through exhibits, seminars, meetings, popular science lectures, workshops, video & scientific film shows and science fairs. It also publishes science magazines, booklets, leaflets, etc.; encourages young scientists to undertake their innovative works and conducts other related activities.

Aims and Objectives

The following objectives have been set forth for the smooth functioning of the National Museum of Science & Technology:

- To popularize science & technology among all levels of public specially school children throughout the country.
- To supplement science education in schools, colleges and other educational institutions.
- To create scientific literacy and awareness among the masses.
- To display the contribution of science and its importance towards the benefit of the human civilization.
- To encourage and help young and non-professional scientists and science club members in scientific and technological innovative works.
- To motivate the public for creating interest towards science & technology and to develop scientific attitude among the masses.
- To inform the people about nature, natural resources and scientific research of Bangladesh and of the world at large.

Functions of the Museum

The NMST at present has been performing the following activities for achieving its objectives:

- Preserves, demonstrates and explains the scientific exhibits in the galleries.
- Supplements the formal teaching of science in schools and colleges in a non-formal way.
- Arranges seminars, discussion meetings, popular science lectures and workshops.
- Plays a vital role in arranging the National Science and Information & Communication Technology Week throughout the country every year.

- Provides library facilities to young scientists, students and educationists.
- Arranges sky observation and mini-planetarium shows for creating interest in space science.
- Arranges display of equipment & instruments innovated by potential young scientists.
- Co-ordinates, encourages and helps science club members in their creative activities.
- Offers the facilities of browsing, e-mail account etc. to the visitors especially to the school children through the Cyber Corner.
- Publishes a quarterly science magazine, popular science lecture series and annual report on the participating projects in the National Science and information & Communication Technology Week.
- Displays the history and chronological development of computer technology through computers in the information technology gallery.
- Promotes a scientific attitude in children's mind through Science Park by way of playing with real objects.
- Develops the science & technology projects innovated by the young and non-professional scientists through financial & technological assistance.
- Provides workshop facilities in developing the projects innovated by the young scientists.

Activities of the NMST

One of the main functions of the Museum is to demonstrate and explain the scientific exhibits displayed in different galleries of the Museum. According to the character/types of exhibits they are at present installed in 6 different galleries namely, Physical Science Gallery, Industrial Technology Gallery, Biological Science Gallery, Information Technology (IT) Gallery, Gallery on Young Scientist's Projects and Science Park. The gallery Assistant in the respective gallery explains the theories and generates science awareness among them.

Museum arranges seminars, discussion meetings, popular science lectures, workshops, video and scientific film shows to supplement science education in schools, colleges and other educational institutions.

The National Museum of Science & Technology has a Cyber Corner to offer the opportunities of generating awareness and interest about contemporary issues from information technology to the new generation.

The National Museum of Science & Technology plays a vital role in arranging the Science and Information & Communication Technology week

A list is given below of the exhibits displayed in different galleries:

Physical Science Gallery

Some of the exhibits displayed in this Gallery are mentioned below:

- Magnetic Lines of Force

- Generation of Electricity
- Generation & Distribution of Electricity
- Electro magnet
- Electric Bell
- Demonstration of Newton's laws of Motion
- Newton's 3rd Law of Motion
- Boyle's Law
- Torricellea's Experiment
- Cinemascope (Nerves Response)
- Static Inertia and Kinetic Inertia

Industrial Technology Gallery

Some of the exhibits displayed in this Gallery are mentioned below:

- Lino Compose Machine
- Mechanical Advantage of Wheel and Wheel Axis
- Mechanical Advantage of Lever
- Barograph
- Deck Watch
- Old Plane Engine
- Wankle Engine
- Old Gramophone
- Echograph
- Antitank Radar and Radar Display
- Life Buoy
- Diving Set
- Loran-C
- Compass
- Old Printing Press
- Mechanical Advantage of pulley
- Master Gyro Compass
- Wireless
- Life Boat Transceiver
- Engine Air Trooper
- Chittagong Still Mill
- Solar Electricity
- Weather Facsimile
- Life Raft

- Wireless
- Fire Extinguisher
- Decca Display
- Model of Karnafuli Paper Mill

Biology Gallery

Brief descriptions of some of the exhibits in this Gallery are:

- Model of Amoeba
- Bony Fish Structure (Model)
- Paramecium (Model)
- RNA or Ribonucleic Acid (Model)
- Model of Cell Division (Meiosis)
- Human Brain (Model)
- Human Heart (Model)
- Human Uterus (Model)
- Respiratory System (Model)
- Insectivorous Plants
- Dionea
- Fossil Replica (Board)
- Human Skull
- Classification of Animals
- Marine Pisces Board
- Bacterial Cell Structure (Model)
- Liver Fluke (Model)
- Hydra (Model)
- DNA or Deoxyribonucleic Acid (Model)
- Model of Cell Division (Mitosis)
- Human Kidney
- Blood Circulatory System (Model)
- Human Digestive System (Model)
- Endocrine Glands (Model)
- Pitcher Plant
- Drausera
- Skeleton of Blue Whale
- Petrified Tree
- Gangetic Dolphin

Information Technology Gallery

In this gallery computers of different era are installed chronologically and being explained to the visitors. The following are the items of the Gallery:

- Analogue Computer
- IBM 1620 System
- IBM 1401 System
- Honeywell DPS-6
- IBM 4331 System
- IBM PC
- Touch Screen Kiosk

Gallery on Young Scientists' Projects

The projects innovated and displayed in the National Science and Information & Communication Technology Weeks, which are developed in the Museum under the supervision of expert guides chosen from different Universities and Science Organisations, are installed in this Gallery. A list of the projects is given below:

- Wood Planer Machine
- Road Barrier
- Magnifying Glass Grinding Machine
- Air Cooler
- Air Cleaner
- Incubator
- "Durber" Modern Van
- Power Pump
- Three Wheeler Car
- Propeller to Control the Boat
- Autovan.

Science Park

In this park at present we have the following exhibits and it will be enriched gradually.

- Levers reduce effort
- Cone runs uphill
- Musical tube
- Archimedean screw
- Lift yourself
- Sympathetic swing

- A seasaw lever
- Swing pendulum.

Future Plan of the Museum

The NMST has a plan for upgrading its function in the near future, undertaking the following activities:

- Establishing divisional branches of the National Museum of Science & Technology outside of the capital, Dhaka.
- Arranging mobile science exhibitions throughout the country through Museo-bus.
- Setting up of a large aquarium.
- Upgrading the existing galleries of the Museum and establishing more new galleries along with an energy gallery.
- Arranging Science Olympiad competition.
- Launching a website.

Co-operation with the Museum

The NMST always invites co-operation and assistance of the interested persons among the general public for enriching the Museum by following means.

- Informing the museum about inventions of the local scientists.
- Donating old and historical scientific apparatus, books etc., or providing relevant information to the Museum.
- Informing the National Museum of Science & Technology of rare animals (living or dead), fossils, minerals etc or providing relevant information.
- Donating products of scientific research organizations and industrial units to enrich museum collections.
- Informing the National Museum of Science & Technology about the science clubs and innovators not yet enlisted in the Museum.
- Helping Museum activities with exhibits, transport, constructive criticism, advice etc.

SPACE THEATRE

Over the past seven decades the planetarium has evolved into an important institution to impart non-formal science education in society. Its role to inculcate scientific temper and to eradicate superstitions and blind faith amongst people has been well recognized in societies the world over. Keeping this in mind the Ministry of Science and Information & Communication Technology recently has established a Space Theatre in Dhaka named Bhasani Novotheatre, Dhaka (Figure 1.1).

This Novotheatre is not only regarded as a classroom of general study, but it also provides scientific knowledge to the student community. It has in its paraphernalia systems such as Astrotec perforated alluminium curtain, GSS-Helios Space Simulator and many other special effects projectors. This Theatre is a unique centre for science education through full entertainment unlike classroom education. This will provide a planet show, which will bring the audience into an imaging dream. Superb images could come up from highly specialized projectors, striking astrovision clips and sky-scan videos will move the audience to a journey to the solar system, which will unfold the ultimate limit to the visible universe and the Milky Way Galaxy. Finally that would offer us the scenarios of death of a star as well as how it would be lost into a Black Hole.

Figure 1.1: Front View of Bhasani Novotheatre, Dhaka (Space Theatre)

Bhasani Novotheatre like other modern space theatres will not only convey the spirit of Astronomy but also offer various networks on Biology, History, Industry, Geography and Anthropology.

The objectives and target of the space theatre are as follows:

- To establish Space Theatre as a non-formal science education institution.
- To inculcate positive scientific attitude in the public mind and to help eradicate superstitions and blind faith from the society.
- To create interest in young generation in studying science & technology.
- To create facilities of recreation for the general public, especially for the students.

Services

The following are the services provided by the space theatre:

- Planet show
- Large format movie on natural resources of the world
- Astronomical exhibits

CONCLUSION

Science centres and museums are playing a major role in engaging the public in a genuine dialogue about science and the issues that it raises for the society. This challenge is an extremely important one if we are to realize the benefits of science. We have to take some steps forward and a long way to go, and much to learn. This is why this workshop is so valuable. It will enable us to exchange ideas and forge new partnerships, and make a critical difference in this important dimension of our lives.

2

The Special Consideration for Establishing Science City in Egypt (Science for All)

Salwa Abdalla El-gharib
Head, Industrial Design Department,
Faculty of Applied Arts, Helwan University, Egypt
Fax: 202-3444057, Email: prof_salwa@hotmail.com

ABSTRACT

In Egypt there are several art, archaeology, history and culture museums. But there are only two small science museums. The state ministry for science and technology as part of its role in raising the scientific awareness, started to prepare for a huge project containing a scientific city, at the level of Egypt's reputation and matching with the modern international systems in this field. A special location was thus chosen carefully at Sixth October city, one of the national capital's new satellite cities, approximately 10 Kms from the Giza pyramids. We were very interested in being informed about any previous experience in similar projects and a vision of how this project could be.

Keywords: Science city; Location; Similar projects

INTRODUCTION

In Egypt there are several art, archaeology, history and culture museums. But there are only two small science museums:

First: Children's Museum

Established in June 1996.

Children from 3 years up to 6 years.

Heliopolis Social services.

Second: Susan Mubarak Science Exploration Center

The first interactive educational science museum in Egypt.

Established in 1998 (Ministry of Education)

The national need for an interactive science museum in Egypt is what Science City is attempting to satisfy. Science City is the first and biggest interactive science city for all ages (Ministry of scientific research).

The initial studies for the project began in May 1996. Dr. Saroj Ghose, Director General of the national council of science museums, India, and president of the international council of museums, Paris, was invited for 2 weeks by UNESCO, Cairo, to advise the Ministry of scientific research in Egypt for planning and developing an interactive science city in Cairo.

A report was prepared at an earlier stage through the preparation committee, consisting of different scientific specialists.

The actual planning, site preparation and architectural design were carried out from 2002 onwards.

Dr. Saroj Ghose was invited again to Egypt (2002) to see the winner architecture and to help on formulating the Terms of Reference with the executive committee and architectural design of the city. From 2002 till now the architectural designer and the exhibit designers have been working on the first phase.

MATERIALS AND METHODS

Architectural Design

Architecture provides the spatial layout for the exhibition of scientific ideas. Optimally the location and connections between exhibits facilitate movement between exhibits of related subjects. This document harmonizes the scientific content of the exhibitions as specified in the Terms of Reference with the architectural design of the city.

(1) Special Consideration of the Project in the Implementation Stage

Location Consideration

- The land allocated to this project lies in 6 October city. It was chosen on the basis of its nearness to the capital especially after the construction of 26 July road.
- The Science City site which reaches 30 acres (126,000 M^2) permits a big area to serve the exhibits and the activities, in addition to a big area to serve the open exhibits, gardens, and parking area.
- It is near the mass media production city as an attractive element.

(2) Special Consideration of the Architectural Planning

The building for this city must be an architectural masterpiece of futuristic characteristics, so that it will be a distinct mark which expresses the inside of the science city.

- We have to keep, as much as possible, the well known ratio in building museums, which is to leave an empty area equal to that of the building (ratio 1:1) *i.e.* if the maximum height of the building is three floors, then you have to leave 2/3 of the land area (landscape) vacant.
- The buildings include wide halls corresponding to the exhibits, and provide room for change and further developments.
- To make it possible to divide the construction work into several stages (4-5 stages) so that the finished stages will not be affected.
- Suitability of the architectural design to some of the exhibits which may occupy big areas or be very tall.
- Representation of the city components in the main building as terminals to move from them to the other buildings with an unusual method (according to the activity) so that you can feel the activity of these building from the specialized area of the main building. The subsidiary buildings can be connected to each other according to the joining relations between the activities.

(3) Consideration Related to the Branches of Science and Technology Represented in the Science City

The necessity of the existence of knowledge and basic sciences in the science city in one way or another, is such that it has to contain the following; the suggested design should keep in view the freedom of movement in the representation of these sciences and their classification:

- Space and its technologies.
- Earth and physical sciences.
- Oceans, seas, and the life of the sea creatures.
- Mathematics.
- Chemistry and its branches.
- Physics and its branches.
- Environmental sciences.
- Energy.
- Future sciences.
- Medical knowledge
- Engineering sciences.

(4) Considerations Related to the Exhibits and Demonstration Methods

- The exhibits represent the most modern achievements of science and technology without neglecting the historical background.

- The exhibits include special objective samples aiming at the complete exploration of all theoretical and practical sides.
- The exhibits must have dynamic characteristics to deal with the current issues continuously, *e.g.* the issues of biotechnology, genetic engineering, remote sensing, advanced ways of imaging, space researches, new materials, informatics, robots, and advanced computer sciences.
- Interaction of the exhibits with different ratios to suit the different types of visitors.
- The exhibits must contain historical models (originals or false), as also live models e.g. the river Nile from source to mouth, putting the big projects built on it or planned to be built (live diorama).
- Open air exhibits such as rare flowers and plants, moving dinosaurs, real oil driller, or dancing water with lights and music exhibition... etc.

(5) Expansion Considerations to Serve the Community

- The Science City must be a productive and distributive center for computer programs and laser discs and video tapes in science and technology subjects.
- Developing models, and copying them for generalization.
- Translating the foreign scientific programs to the Arabic language in cooperation with the publishers and governments.
- Giving lectures and presenting demonstrations in the current scientific subjects, by Egyptian and international speakers.
- Providing services for scientific seminars and conferences carried out in the city.
- Marketing of scientific research outputs and linking them to industry needs.
- A scientific library of high quality allows reading and research services.
- Extending services to get to cover other governorates.

Science Park Activities

The park is designed with nature and technology as two ends of the spectrum of human experience. The park is a landscape of discovery and imagination. A microcosm of the potentials of nature and ingenuity of humankind.

All activity areas are connected through well marked pathways arranged in a functional aesthetic manner avoiding harsh straight lines except for key entry points. The flow of visitors is from four key points with a central point. A ring road (T-track) along the outside periphery provides fast access. Another fast track runs from the northern corner to the center.

ACKNOWLEDGMENT

We thank Prof. Dr. Fawzi A. El-Refai, President of the academy of scientific research and technology (ASET), and Prof. Dr. A. Hebaish, chairman of the executive committee of the Egypt Science City for all their support to the project. The contents of the work are based on an architectural design by Gamal Bakry and Dr. Fekry Hassan.

3

The Changing Role of Science Centres and Museums in Developing Countries: An Appraisal

Samir Kumar Ray
Director, Central Research & Training Laboratory (CRTL), Block-GN, Sector-V, Bidhan Nagar, Kolkata-700091, India
Phone: +033 23579347-48(O), 23573937 (R), Fax: +033 23576008, E-mail: ncsmin@vsnl.com, samirray52@yahoo.co.in

ABSTRACT

In today's era of information technology there is a growing awareness of the need to equip our populace with scientific learning and thinking skills. This paper outlines various features and importance of imparting non-formal education for popularization of science among the populace. The ability of a science museum/centre to elicit a sense of excitement, of wonder and to provide concrete experience through involvement, discovery, fun and adventure (?) may be considered as some of the parameters by which we can judge the success of such an institution. Science museums/centers should be like open universities, to a certain extent. They must dedicate their resources to educate the populace about scientific phenomena and implications of technological advancements in an entertaining way through varieties of approaches with an objective of dissemination and popularization of science, inculcating scientific temper and eradicating scientific illiteracy thereby making this information age more comprehensible to one and all. However, it must be appreciated that the needs of a citizen from a developing country are different from the needs of his/her counterpart in a developed country and such issues have to be attended to accordingly.

INTRODUCTION

The World Conference on "Science for the Twenty-First Century: A New Commitment" organized by UNESCO and the International Council of Scientific Unions in June 1999, held in Budapest, in its Declaration on Science & the Use of Scientific Knowledge proclaimed:

"Science education, in the broad sense, without discrimination and encompassing all levels and modalities is a fundamental prerequisite for democracy and for ensuring sustainable development... It is more than ever necessary to develop and expand science literacy in all cultures and sectors of society as well as reasoning ability and skills and an appreciation of ethical values, so as to improve public participation in decision-making related to the application of new knowledge."

UNESCO's commitment as above indicates clearly that in the global scenario the role of science museums/centers is unique: to carry the message of science to one and all, through the non-formal approach of science education, as the formal educational approach is not sufficient to meet the curiosity of the populace.

SCIENCE & TECHNOLOGY EDUCATION IN DEVELOPING COUNTRIES

The scenario of science education in many developing countries across the globe is quite disappointing at present. Fund crunch resulting in lack of required basic infrastructure support has led the basic science education system in these countries to be dependent mostly on theoretical teaching i.e. the empty vessel approach with chalk and talk. This in turn takes away the thrill, fun and enjoyment of learning science and as a consequence there is insufficient appreciation of the present day progress in scientific and technological endeavors.

For true prosperity of a nation the right step is economic development no doubt but can we achieve this by keeping proper education of the populace in isolation? In fact it must be appreciated that if the aimed socio-economic development is considered as analogous to a metal cutting machining job to be done, then the role of literacy should be treated as the cutting tool. What is necessary is to take appropriate action to spread the required functional literacy so that the countrymen can take part actively in this massive endeavor. Once we are able to achieve this it will also help abandoning superstition, obsolete concepts and scientific thinking will replace orthodox authority.

The Theme *"Science & Technology for all"* should be the keyword in our present day society. Society provides the institutions within which the professionals work. Human values, needs and problems determine in large part what questions professionals investigate. In this process new scientific knowledge and technologies bring about changes in human society.

SCIENCE MUSEUMS/CENTRES

A few decades had elapsed prior to the emergence of the present day hands-on science centers (**activity oriented**) concept around 1970s from the science museums (**object oriented**) concept. However, both science museums/centers have a distinct characteristic (compared to art and archaeological museums etc.) of being interactive science & technology-based institutions imparting non-formal science education.

The Language of Communication

In science museums/centers, the language used for communication is **exhibits**. Besides, educational programmes/outreach programmes etc. cater to millions of visitors. Here, learning being through play, fun and interactivity, it leaves a permanent impression. Here what is complicated within the four walls of a class room is made simple and comprehensible. In these institutions the visitors (learners?) are the clients and their interactive participation is of vital importance as it ensures success of the exhibits and activities.

THE CHALLENGE

In today's era of globalization where success in the cutthroat competition mode has emerged as the key to survival, the changing role of science museums/centers particularly in the developing countries is basically to meet the challenges being faced due to:

- Changing technological scenario.
- Changing social scenario &
- Changing global scenario.

In order to remain competitive, we must be able to adopt the technological change. Delay, indifference or inertia will bring disaster. Technological changes are taking place fast and new technologies are advancing rapidly. Globalization of economy is now a reality poised to stay and grow. Quality rather than price is gradually emerging as the preference of our populace. There is an increasing concern on ecological and environmental issues in the global perspective and the demand for **eco-friendly** technologies is on the rise. Add to this the fact that the technological life cycles are shortening which increases pressure on competition. We have to meet this challenge with appropriate approaches.

To be successful in such a situation we have to consider the following for implementation:

- Development of suitably trained manpower.
- Development of a sound research base to augment indigenous as well as appropriate technology with active participation of our populace.

We have to plan and act in a way so as to draw out the creative potential of our own countrymen. Managing this gigantic task properly and, in the process, to come out successful is definitely a cultural challenge and an economic compulsion as well.

Our goal should be "Taking the message of Science & Technology to one and all" and this is the right direction for a developing country to progress to become a developed country in the days to come.

In this context, it will definitely be of immense value if we make **SWOT** (Strength, Weakness, Opportunities, Threats) analysis of our own standing vis-à-vis the existing scenario related to the Science & Technology development in our countries and plan appropriate strategies to effectively face the challenge of global competitiveness.

Successful implementation of the above will ensure adequate support base to achieve results mostly related to:

- Increased productivity and quality.
- Increased generation of revenue.
- Cost minimization/optimization.
- Acquiring self-sufficiency.

Achieving **excellence** should be our aim (Figure 3.1).

To achieve enhanced performance of our human resources, development of the organization along with its workforce has to be judiciously planned and implemented. Development and training must go hand in hand.

Training of human resources should be considered as **an investment** and it should be a continuing process backed by appropriate resources including a need-based budget.

The scope of developing our human resources should encompass steps to:

- Impart knowledge.
- Develop skills.
- Ensure change in attitude.
- Improve motivation.

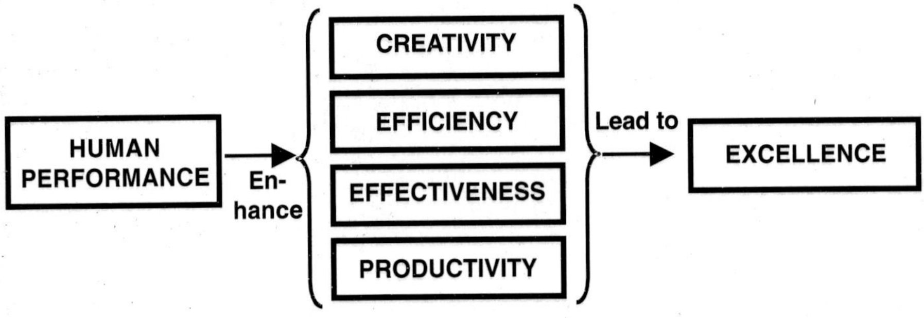

Figure 3.1

The Changing Role

Considering the above requirements, imparting education through the non-formal approach becomes most suitable and appropriate and to achieve this the role of Science Museums/Centres has to be unique. Science Museums/Centres should be made to function like *Open Universities* to impart non-formal science education to our populace. They should dedicate their resources to bring scientific awareness among the populace with the following basic objectives.

- To supplement the formal education imparted in our schools and colleges.
- To organize suitable programmes to inform the populace about scientific phenomena and implications of technological developments in an entertaining way through a variety of approaches.
- To generate a scientific culture in the present knowledge based society.

In this context, it is pertinent to mention what Pandit Jawaharlal Nehru, the first Prime Minister of India had said: *"It is science alone that can solve the problem of hunger and poverty... The future belongs to science and to those who make friends with science..."*.

CONCLUSION

The Kolkata Declaration issued after the successful completion of the international workshop on "The Role of Science Centres in Development Strategies" in 2002 proclaimed among others:

- There is a need to establish a *network* among science centers, museums and other institutions engaged in science popularization in the developing countries for effective sharing of individual experiences...
- Extensive science popularization activities particularly in the rural areas in the developing countries are required...
- The science centers in the developing countries should endeavor to include the *traditional knowledge and innovation* as well as the significant landmarks in the development of S&T in their respective countries as part of their exhibits.

The situation has not changed much during the intervening period and it is felt necessary that the developing countries must work in unison to achieve the goal.

We have a long way to go to achieve our goal and we must succeed.

REFERENCES

1. "Science, Technology & Industrial Development in India", a Publication of Publication & Information Directorate, New Delhi-110012 and Wiley Eastern Ltd., New Delhi-110002, 1995.
2. Museum International, Pp.3 No.4, Oct.-Dec. 2000
3. Mukhopadhyay, Ingit K. 2002, "The Role of Science Centres in Development Strategies" Keynote address delivered in the international workshop on the Role of Science Centres in Development Strategies, Kolkata.

4. Sneider, C. et al, 2002 "Technology and Its Makers" A New Perspective for Science Museums, ASTC Dimensions, Sept.-Oct. 2002 Pp.6.

5. Ray, Samir Kumar, 1993 "Science & Technology Policy, Scientific Illiteracy and role of Science Museums/Centres", Proceedings of the Seventh Indian Engineering Congress, Bangalore, India, Pp. 82-89.

6. Ray, Samir Kumar, 2003 "Training of Science Museum/Centre Professionals: Activities & Issues", Paper presented at the Third National Conference of the Directors' of Science Museums/Centres, Pilani, India.

4

A Presentation on Educational Programmes and Activities Organised by National Council of Science Museums, Govt. of India, and A Live Demonstration Lecture on Physics: Fun with Papers

Ratan Kumar Chakrabarti
Education Officer 'C'
Birla Industrial and Technological Museum, (NCSM),
Government of India, 19A, Gurusaday Road, Kolkata- 700019, West Bengal, India
Phone: +91(033) 2247-7241/7242/7243, Fax: +91(033) 22476102
E-Mail: bitm@cal2.vsnl.net.in

ABSTRACT

The science centres and museums under the National Council of Science Museums located in four zones of India [East (at Kolkata, BITM), West (at Mumbai, NSC), North (at Delhi, NSC) and South (at Bangalore, VITM)] along with its State and District Science Centres/Museums and Science Parks organize and conduct different types of Educational Programmes and Activities in every corner of our country throughout the year. The target groups are common people (men & women), literate or illiterate, children, parents and, in particular, school and college students. The main objectives of our museums are popularizing science and technology and to promote scientific awareness among all classes of people of our country. Included are glimpses of educational programmes organised and conducted throughout the year by the Science Centres and Museums [under N.C.S.M] along with the State and District Science Centres.

INTRODUCTION

We know that civilisation of mankind and development of science, technology and society and their interaction produced a free spirit towards innovation, which assisted to remove the fetters of old customs and encourage the acceptance of new ideas of invention and application of modern technology in different fields.

All that man knows is that he learns through his own experience and interaction. These two demand, a two-way communication with the messages transacted back and forth. When an educator in our museums appears on a stage or in a classroom for delivering a lecture or demonstration, initially he has a visual appeal with or without a message. When he starts speaking he has an additional appeal through his language and presentation of the experiments in front of the audience. Audience may be common people, students or children who then start interacting with the lecturer or demonstrator. This is the way a two-way communication is established between the audiences and the demonstrator and the whole demonstration lecture becomes lively, attractive and easily understandable.

There is a meaningful proverb "I read and I forget", "I see and I remember", "I do and I understand".

Learning science by doing experiments with pleasure and enjoyment is one of the ways of doing our educational programmes and activities. Science Centres and Museums are lifetime learning places, the hubs of non-formal or rather informal science education in India.

On the other hand explaining without props or hands on/minds on approach is like teaching swimming without letting a person enter the water. Thus there is a growing need to create public understanding of science through a process, which is not only comprehensible and enjoyable but which facilitates public participation, allowing participants to handle equipment and experiments as well.

These Science Centres and Museums under NCSM in India offer a variety of opportunities to participate in different educational programmes and understanding scientific concepts and interest development.

These programmes and activities foster curiosity and creativity among the children and common people. The fostering of curiosity is the first step forward in bridging the gap that exists between the existing levels of scientific knowledge of the common people and that, which is necessary to comprehend the tremendous advances taking place in different areas of science and technology. Popularisation of science is as important as creating new scientific knowledge, for it promotes science education in a non-formal mode and, thus, plays an instrumental role in raising scientific consciousness among the general public. Science Museums have proved to be excellent learning centres that enhance the public understanding of science.

The 28 Science Centres/Museums under the National Council of Science Museums in India are not mere storehouses of scientific curios but are living scientific laboratories where everyone has freedom to explore, experience and experiment.

FUN WITH PAPERS

A Demonstration with Experiments

This demonstration can be performed with a few pieces of ordinary papers. (This will be an example of the science demonstration lectures we perform for school students.)

List of the Experiments to be Performed

- Stunt experiment: paper that does not burn. (An experiment to explain what is latent heat).
- Tearing a piece of paper slowly and rapidly. (To show the difference in the frequency of sound produced in each case for explaining that frequency is directly proportional to the pitch of the sound).
- Why is it easier to tear the paper along a fold?
- A corrugated piece of paper can bear a weight but not a plain paper of the same size. (To explain how the forces are distributed in different directions).
- To simulate the blowing off of a house roof (paper made). (To demonstrate Bernouilli's principle).
- Helmotz's resonator with paper cones of different sizes. (To explain how the resonance of sound occurs in paper cones of different sizes).
- Blooming of paper flowers in water. (To demonstrate the capillary action of liquid flowing in narrow capillary tubes).
- Breaking of a wooden scale with a newspaper. (To demonstrate how strongly the atmospheric pressure works).
- Preparing hot coffee in a paper-made pot. (To explain the difference in conductivity of heat through different media).
- Tracing any drawing using an oily paper. (To explain properties of reflection and refraction of light).
- Why we can write on paper but not on glass with an ordinary pencil. (To demonstrate the important role of friction in our day to day work).
- To demonstrate inertia of rest using a piece of paper and a weight.
- Playing paper flutes of different lengths. (To demonstrate the relation of sound pitch on the length of the vibrating air column).

[Answer of the Stunt Experiment]

The piece of paper is dipped in a mixture of water and alcohol and then held on a flame. The alcohol starts burning and the water keeps the paper cool. For burning the latent heat is taken from water not from the paper, so the paper does not burn.

ACKNOWLEDGEMENTS

1. Centre for Science and Technology of the Non-Aligned and Other Developing Countries (NAM S&T Centre), New Delhi, India.
2. Ministry of Science and Technology (MOST), Centre for Regional Research and Development (CRD), Hanoi, Vietnam.

Abbreviations

BITM: Birla Industrial And Technological Museum, Kolkata.

NCSM: National Council Of Science Museums, Kolkata.

NSC: National Science Centre, Delhi.

NSC: Nehru Science Centre, Mumbai.

VITM: Visvesvaraya Industrial and Technological Museum, Bangalore.

5

Changing Role of Science Centres and Museums in India

V.B. Kamble
Director, Vigyan Prasar,
Department of Science and Technology,
New Mehraulli Road, New Delhi–110 016, India
Phone #: +91-11-26864022, 26967532 Fax # +91-11-26965986
E-mail: vbkamble@alpha.nic.in

SCIENCE CENTRES AND MUSEUMS: THE BASIC RESPONSIBILITY

If one were to state the aim of a science centre or a museum, it could be "to inculcate scientific temper". It is difficult to give a precise definition of scientific temper; however, fortunately, it is not necessary to do so. Broadly speaking, scientific temper is an attitude of mind, a way of life, a manner of viewing facts and information. On an individual level, it forms an integral part of one's personality. In the context of a society, it decides how the community will approach any given problem and leads to the evolution of value judgments the community would cherish. In short, it is an attitude that shapes patterns of individual and group behaviour. In that sense, scientific temper transcends various branches of physical and social sciences and technology. Obviously, it is more than a mere collection of facts or acquisition of knowledge; rather it concerns the wisdom with which the facts are seen and interpreted and knowledge utilised. There is no gainsaying the fact that spirit of enquiry does not relate only to scientific matters, but, it equally permeates other spheres of human activity as well, say, language, society, sports, administration and so on.

The major function of science centres and museums is to supplement the formal education imparted in schools and colleges. One encounters science in every walk of life, say, in agriculture, in medicine, in nursing, in child care, in mitigating disasters and so on. It is important to realize how various disciplines of science come together in an integrated manner when we observe a natural phenomenon. Even in the social context, while dealing with problems and issues related to health, environment, water, energy, and so on, one needs to consider physical, chemical and biological aspects simultaneously. It is hence imperative that a holistic and integrated view of the phenomena we observe in our environment–both physical and social–is given. This will help people appreciate the intimate relationship of science with their lives. It is in this sense that science centres and museums are facility-cum-resource centres.

SCIENCE COMMUNICATION IN INDIA: PRESENT SCENARIO

With the impact of science and technology (S&T) on social life and economic development, the importance of S&T popularization was realized decades ago in India. With the establishment of governmental agencies like the National Council for Science and Technology Communication (NCSTC) and Vigyan Prasar, and other voluntary agencies coming into being in the last twenty five years or so, a variety of activities now take place all over the country. It is gratifying to note that agencies like the National Council of Science Museums (NCSM) and National Council for Educational Research and Training (NCERT) have been contributing to the field of S&T popularisation significantly. NCSM has a chain of 28 science centres and museums at district, regional and national levels with 300 affiliated school clubs. Vigyan Prasar has a network of nearly 6,000 clubs spread throughout the country. NCERT has a mandate of developing curricula and preparing text books at school level, and training and research in the field of education. NCERT also organises programmes like science exhibitions for schools. Besides, there are a few science centres such as the Homi Bhabha Centre for Science Education, Mumbai; Vikram A. Sarabhai Community Science Centre, Ahmedabad; and Vidnyan Ashram in Pune, engaged in non-formal science education and popularisation of S&T. There are 15 state-of-the-art planetaria spread in different parts of the country. A large network of centres for awareness about modern agricultural practices–also called Krishi Vigyan Kendras–under the Ministry of Agriculture, serves farmers and their needs. S&T coverage in newspapers/magazines is also steadily picking up.

As regards the media, All India Radio broadcasts at least one programme everyday on a scientific topic from all its stations, in addition to the ones on health and agriculture. Several AIR stations with science cells broadcast three programmes per day. Doordarshan–the National Television Channel–telecasts about two programmes every week on the national network and all regional centres put together produce and telecast some 150 S&T programmes every year. Agencies like the University Grants Commission, The Central Institute of Educational Technology, and The Indira Gandhi National Open University also have regular slots on Doordarshan. Programmes on S&T popularization for telecast/broadcast are also produced by several Government/Non-Government agencies. NCSTC has been encouraging publication of popular science books/magazines in regional languages besides utilizing various media and means and networking of different peoples' science groups.

Pioneering Efforts

The Community Science Centre established at Ahmedabad in 1966 by the visionary scientist, Dr. Vikram A. Sarabhai, the Father of the Indian space programme, and the NCSM set up in the mid-seventies, have been the fountainheads of the pioneering efforts of science communication and the public understanding of science in India. The need for public understanding of science was beautifully brought out by Dr. Vikram A. Sarabhai at the time of the foundation laying ceremony of the Community Science Centre (Box 1). In particular, "Science Playground" is a unique concept that travelled from the tiny Community Science Centre to the West through NCSM (Box 2).

The National Children's Science Congress is being organised by the NCSTC-Network–a voluntary association of over sixty non-government, government and voluntary organisations–for the last twelve years in the country and has already caught the imagination of a few western and west Asian countries. In this programme some 600,000 school children participate. They undertake studies/projects on specific scientific themes. The reports are then presented at school, district, state and national level congresses. The winners then participate in the special session of the annual Indian Science Congress in January every year. School children from a few other countries participate in this event on a regular basis. National Childrens' Science Congress; S&T popularisation activities built around the total solar eclipses of October 24, 1995 and August 11, 1999; Venus transit of 08 June 2004; Emergence of Modern Science; S&T for the visually challenged; teaching/learning aids and toys; networking of Government/Non-Government organisations for S&T communication, are all pioneering efforts in S&T communication. Bharat Jan Vigyan Jatha 1987 and Bharat

BOX 1

NEED FOR PUBLIC UNDERSTANDING OF SCIENCE

The development of a nation is intimately linked with the understanding and application of science and technology by its people... and ability to question basic assumptions in any situation is fostered by the frontiers of science... It is this ability rather than an empirical hit-and-miss approach which proves most effective in tackling the day-to-day problems of the world.

A broad understanding of the physical and social environment in which one lives is the most urgent task which faces all humanity... Prejudice and a feeling of omnipotence permeates the minds of those who without bothering to understand science enjoy the rich rewards flowing from the application of technology. Today, the consequences of failure to raise the level of enlightenment are more serious to the security of the world than they ever were before. The task of promoting and understanding of science is, of course, at the core of the problem of education and becomes increasingly more difficult in the context of a population explosion.

— **Vikram A. Sarabhai**

BOX 2

WHO THOUGHT UP THE IDEA OF A SCIENCE PLAYGROUND?

In 1980 I visited India to watch a spectacular solar eclipse as the moon's shadow raced across the subcontinent. After the excitement of the eclipse, my wife and I did what most museum staff do on vacation–we visited museums. India has long had a wealth of important art and history museums, and by 1980 it had already developed a superb network of hands-on science museums. Along with our own network in the U.S., the Indian network was one of the best organized in the world.

One of the Indian museums, a small science center in Ahmedabad called the Vikram A. Sarabhai Community Science Center, had invented a new exhibition concept: a "science playground". The outdoor space around the Community Science Center was filled with large scale, imaginative science demonstrations. Putting exhibits outdoors saved the cost of constructing a building to house them. Even more important, the outdoor exhibits had a scale and playfulness that encouraged visitors to engage and explore the apparatus in an especially joyous manner.

Because the Indian museum network functioned so well, science playgrounds were under construction or in planning across the country, encouraged by the National Council of Science Museums of India. I visited India again in 1982 and saw large science playgrounds in Bombay and in Patna. There were big pulleys and weights to lift, special swings and family-size seesaws to ride, and whole-body spinning chairs which let visitors experience angular momentum in an unforgettable (if dizzying) fashion.

Within a few years American science centers learned about the Indian science playground concept. A couple of museums here had also independently come up with the same general idea. The St. Louis Science Center built the first major science playground in this country. When the New York Hall of Science opened in 1986, our expansion plans already included a major science playground of our own. Our exhibits include some inspired by the Indian playgrounds, some imported from Europe, and some conceived by our own staff and consultants.

The new science playground at the Hall of Science is a great example of how ideas are shared and enriched by museum collaboration around the world. See you soon, on the giant seesaw!

— Alan J. Friedman, Ph.D.
Director
New York Hall of Science
(From the website of New York Hall of Science)

Jan Gyan Vigyan Jatha 1992–groups of science communicators–moving in different parts of India spreading science awareness interacted directly with people from nearly 400 districts covering about a fourth of India's population.

Mention also may be made about the Vigyan Prasar Network (VIPNET) of Science Clubs–mostly in rural areas, with nearly 6,000 member clubs as of today. Attempts are being made to connect the clusters of the VIPNET science clubs through the satellite communication network. This has laid the foundations of a national science club movement. A countrywide network of Eco-clubs established to spread awareness about conservation of environment, biodiversity and sustainable development has also been established by the Ministry of Environment and Forests.

As part of the Year of Scientific Awareness (YSA)–2004 programme, a unique effort was undertaken to take science to the people in a train called the "Vigyan Rail" or Science Exhibition on Wheels that moved throughout India for about eight months from December 2003 to August 2004. During its run, it travelled to 60 destinations covering 15,000 kilometres, the number of visitors being nearly 5 million. It had working models, interactive exhibits, computer games and quizzes, charts, scientific videos and so on. The exhibition was the result of cooperation and collaboration between 18 Ministries/Departments of Government of India engaged in various scientific fields. Incidentally, constituent units of NCSM, science centres, R&D labs and a number of Government and non-Government agencies participated in this gigantic effort.

Government Policy

The Science and Technology Policy Statement of Government of India 2003 lays considerable emphasis on the need for popularization of science. It states that: "There is growing need to enhance public awareness of the importance of science and technology in everyday life, and the directions where science and technology is taking us. People must be able to consider the implications of emerging science and technology options in areas which impinge directly upon their lives, including the ethical and moral, legal, social and economic aspects. In recent years, advances in biotechnology and information technology have dramatically increased public interest in technology options in wide ranging areas. Scientific work and policies arising from these have to be highly transparent and widely understood."

The spirit of enquiry, quest for knowledge, need to understand, manipulate, change, seek answers, explain the reasons are probably innate to human nature. The human mind is always in search of answers from simple to complex issues. The search for knowledge is innate to human nature and needs to be guided. At the same time, basic skills also need to be imparted. This can be done in a variety of ways. It can happen through the traditional methods of teaching and learning, it can happen through informal learning, it can happen at the individual's initiative or it can be done by providing the right type of atmosphere and ambience. The science centres and museums are ideally suited to create this ambience. Just as every human activity needs an atmosphere, motivation and stimulation, the scientific quest of knowledge is also guided by this environment. The stimulation and excitement provided by the science centres and museums can make this into a joyous learning experience.

SCIENCE CENTRES AND MUSEUMS: THE CHALLENGES AHEAD

Literacy in India has significantly gone up over the years. Nearly 65% of the people can read and write today, compared to 52% in 1991. Nearly 76% of the males and 54% of females are literate today; in 1991 their percentage stood at 64 and 39 respectively. This implies that more than three-fourths of the males and more than half of the females can read and write today. Our food production has tripled since independence, *i.e.* since 1947. We make most of our drugs and pharmaceuticals, satellites, TV sets and telephones–and many other things. Still, half of our people do not have safe drinking water and nearly 90 per cent of our people are without sanitation facilities. Given 16 major language groups and over 4,000 dialects, the variety of languages spoken, poverty, urban-rural and cultural divide, and disparities in terms of opportunities and facilities, we face challenges which no other country does.

New technologies are fast replacing the "what-were-once-new" technologies as they become obsolete. How fast the "latest" model of a computer purchased by you becomes obsolete, or how fast compact fluorescent lamps are replacing the good old tube lights. Well, this is how it should be. In the field of medicine, terms like ultra-sonography, magnetic resonance imaging (MRI), organ transplant, and in-vitro fertilization have almost become a part of everyday life. Consider the advances in the field of biotechnology. Genetically modified food products, transgenic crops, cloning, gene replacement therapy, stem cell harvesting and similar jargon is fast becoming a part of our everyday vocabulary.

Undoubtedly, the new technologies have helped improve the quality of life, thereby significantly changing our life-styles. However, it is also equally true that breakthroughs in the fields like biotechnology have instilled a sense of awe and fear–fear of the unknown–in our minds. Consider the questions: Are genetically modified food products safe to consume? Will transgenic plants with built-in pesticides give rise to a new breed of pests which are resistant to these built-in pesticides? How shall we tackle the problem then? What is the guarantee that gene replacement therapy would be used only for the human good? Indeed, most of us are totally ignorant of what the new science holds in store for us. Such doubts and questions make the task of realizing our dream–even in a small part– of equpping every citizen with a scientific outlook appear to be a distant dream indeed!

There has been a marked change in the social fabric over the last decade as a combined result of the liberalization in economy and the phenomenal growth in the field of information technology. This process of development has its own associated problems. More and more students have been opting for professional courses and fewer and fewer for a career in pure science, which could result into an acute shortage of scientific manpower in the near future. This imbalance will need to be corrected, since science and technology need to go hand in hand for an all round growth of a society. It is here that science centres and museums can make a difference. Active participation of science centres and museums contiguous with programmes of NCSTC like the National Children's Science Congress, science exhibitions, Emergence of

Modern Science, contact programmes for school and college students with scientists, and visits to R&D laboratories, would go a long way in explaining to them the thrill and challenge offered by a career in pure science.

In India, majority of S&T popularization software and programmes are produced by Governmental agencies. However, in view of the marked change in the social fabric over the last decade, as a combined result of liberalization in economy, the phenomenal growth in the field of information technology, and emphasis on literacy, the expectations and the needs of the people have changed significantly. Further, different sections of the society have markedly different requirements. As a result, the programmes and software to be produced need a good deal of thought and preparation. The different sections of society that would need to be specifically addressed are: School children at primary, secondary and tertiary levels; out of school children; women with emphasis on girl-child and housewives; adult illiterates; employees/workers/artisans; farmers and agriculturists; and college/university students. Indeed, it is a mammoth task to address all the groups, since the software and efforts required for different groups listed above would need to be tailor-made. This would require intensive research in determining their needs and expectations–both in urban and rural areas.

We are on the brink of exciting developments in science that will affect everyone's life. Along with the benefits come potential risks which cause concern. People rightly expect proper safeguards against risks, and full and open discussion about the values, benefits and risks of scientific developments.

HOW SCIENCE CENTRES CAN HELP INFLUENCE PUBLIC OPINION

Scientific attitude does not imply imparting or acquiring information alone. It also implies organizing, analyzing and applying the same in order to arrive at a solution or creating meaningful patterns leading to a systematic understanding of the natural processes or the understanding of the environment we live in–physical or social. For the development and growth of a society, the spirit of enquiry and scientific attitude need to be inculcated at every level in every system, whether at home, office or street–from the lowest to the highest functionary, from a peon to the top bureaucrat, from a child to parents and teachers.

Science being a process and method rather than a mere branch of knowledge, it naturally permeates every human activity. Many of us–scientists, science communicators, social workers, students, teachers, administrators, etc.–have been actively engaged in a variety of societal problems in a bid to transform our country into a nation of a literate and scientifically minded people through various media and modes. There is no gainsaying the fact that science centres have enormous impact in influencing public opinion. Whatever the topic and whatever the medium of communication, scientific approach must form an integral part of our thought and action. Science centres and museums can help in speeding up the transformation of a developing country into a nation of literate and scientifically minded people.

Science is not confined to textbooks, laboratories and is certainly not the preserve of scientists. The human mind seeks to understand, explore, manipulate and analyse the immediate environment. Science is everywhere and for everyone. The basic spirit and method of science are simple, logical, rational and understandable for everyone. Questioning existing explanations and beliefs–be it about a natural phenomenon or a social tradition–and search for truth is the basic aim of science. One may look for an accurate description, explanation or cause and effect relationship in an observed phenomenon. Two examples of how science centres helped promote the spirit of inquiry are described in Box 3.

The socio-economic-cultural conditions have a strong bearing on science. Both the East and the West have had their scientific traditions. There has been a great deal of cross-cultural flow between the oriental and the occidental. India and many other developing countries have had a rich tradition of science and technology. Indeed, there has been significant give and take between both and both have influenced each other. It is hence imperative to inculcate the universal nature of science–a challenge by itself.

Box 3

SCIENCE CENTRES PROMOTE SCIENTIFIC SPIRIT

Only a few years ago in a small nondescript village of Madhya Pradesh, a State in Central India, there appeared white snakelike lines on the green leaves of snake gourd, a popular vegetable grown in many parts of the State. The appearance of these lines was interpreted as the curse of the snake god, and as a result, people altogether stopped eating the accursed vegetable! Curious to know the cause of the appearance of white lines, a few young girls set out to unravel the mystery with the help of their simple plastic microscopes given by a local science centre. The mystery was soon resolved when they discovered that the appearance of the white snakelike lines on the green leaves was due to a tiny worm only about a millimeter in size; and not due to the curse of the snake god.

In yet another example, the school children of the city of Gwalior, also in Madhya Pradesh, took up a survey to find the pollution levels of water supplied by the municipality in different parts of the city using a simple water testing kit under the guidance of Science Centre, Gwalior. They discovered that water supplied in most of the areas was polluted and posed a serious threat to public health. When this report was published in the news papers, the municipal authorities were compelled to take measures to improve the quality of water supplied.

The first example illustrates the basic curiosity we all possess to understand a phenomenon. The second example illustrates a systematic study of a specific problem affecting the entire population of a city. Both the examples show how we can systematically approach a problem and try to arrive at a solution. Indeed, both the examples illustrate what we call the method of enquiry, or the scientific method and the role science centres can play in promoting scientific spirit.

SCIENTIFIC LITERACY

I would like to discuss a related issue–what we call scientific literacy. When we read a newspaper, we can follow with ease, understand, and interpret news items related to education, sports, and politics, and so on. If we can understand with the same ease news items/articles on topics like the launch of a polar or a geostationary satellite, genetically modified foods, use of CNG for transport, anthrax as a biological weapon and so on, it could be said that we are scientifically literate. It is then that one can actively participate in the decision making process. This, however, implies information on and understanding of recent scientific developments and working knowledge of gadgets we come across in our daily lives. For example, information about AIDS or Dengue and how they spread could help us keep them at bay. Knowledge and information about ORS (oral re-hydration solution) could help save lives of thousands of infants. The challenge before science centres and museums is then imparting information on and interpreting such items and issues in a language and manner acceptable to the people.

SCIENCE CENTRES AND PEOPLES' SCIENCE MOVEMENT

However, for the development and growth of a society, the spirit of enquiry and scientific attitude need to be inculcated at every level and every system rather than mere dissemination of scientific information. It may be stated that the literacy campaigns and S&T communication activities need to go hand in hand in order that the skills achieved by the neo-literates are not lost. The skills achieved by neo-literates would be lost unless utilised to understand and appreciate the issues directly affecting their lives; say health, environment, sanitation and hygiene; nutrition, appropriate technology, superstitions and their scientific explanation; means of income generation and the like; and that too without any conscious efforts to inculcate a scientific temper among them. Further, peoples' participation is imperative in a decision making process, say in the formulation of a policy, formulation and execution of development oriented projects etc. With this realisation, a strong peoples' science movement is steadily building up in India through programmes like literacy campaigns, S&T communication programmes built around the events of the Total Solar Eclipses of 1995 and 1999 and so on. Bharat Jan Vigyan Jatha (1987) gave rise to the All India Peoples' Science Network which is wholly a peoples' initiative. It comprises of some 40 voluntary organizations working in the field of S&T communication. What is important to note is that a substantial contribution–both in cash and kind–comes from the people themselves in such programmes. Surely, a good deal of S&T popularization work is going on in India at present, but still there are miles to go. However, our goals of total literacy and a country with scientifically thinking people do appear achievable in view of the peoples' science movement slowly taking shape.

SCIENCE CENTRES AND MUSEUMS: NUCLEI FOR TRANSFORMATION

We have attempted in this article to outline the ongoing efforts in India to communicate science to the people, help them cultivate a scientific outlook and involve them in decision making processes. True, a large number of voluntary organizations are involved in this mammoth task, but the initiative of the Government and support–financial and organizational–cannot be overemphasized. Donations from private sector/individuals are equally important.

Indeed, over the years, the distinction between a science centre and a museum is slowly getting blurred in terms of activities. With growth in literacy in the developing countries, changing socioeconomic scenario, and the revolution in the communication and information technology have made science centres and the science museums to redefine their roles. A museum is no longer a place with a mere collection of items and exhibits, or a science centre with a few activities for children. Their prime responsibility is to make people scientifically literate, trigger an interest in science among them–especially the young, and make them understand that science is everywhere and for everyone.

The new role the science centres and museums have to assume today is to act as a bridge between the people and the R&D labs from where new technologies flow and explain their possible impact on the environment and society. This would help people make informed choices and become scientifically literate. Science centres and science museums are the nuclei where begins a process of transformation of a country into a nation of scientifically minded people.

6

Science Centre Challenge in a Changing World

Herudi Kartowisastro,
Director, Indonesian Science Center, Jakarta, Indonesia
Fax: + 62 21 3101728/3102368
E-mail: herudi-k@cbn.net.id, piaszz@yahoo.com

ABSTRACT

Since the mid sixties Science Center mission in most developed countries is to emphasize on awareness of science and technology for the society and has been followed by many developing countries. But a rapidly changing science and technology scenario has influenced deeply the societies of developing countries. Whether you like it or not Science Center has to adapt to these changes. This changing role has to be in line with the progress of society, development of its culture and also of science, technology and education policy of the country concerned. If these three variables are incorporated in the role of Science Center, the effectiveness of its mission will be more meaningful.The simplest measure of the success of a Science Center is the number of visitors. From our experience, taking into accout psychographics analysis based on the three above mentioned variables has resulted in an increase in visitors. Eighty percent of our visitors are elementary and high school students and their number has increased significantly after adapting the Science Center role to the new government education policy of "Competence Based Curriculum". Our Science Center becomes a "RESOURCE" to schools, as complementary to their laboratories and learning activities. Our interactive exhibits and programs are geared to be more contextual to balance the influence on science and technology of the country. Indonesia as a country with more than 200 million people and inhabited islands is striving for equity and this has made our Science Center as an agent of development of more Science Centers in major cities throughout the country. Two to three times a year we run our OUTREACH PROGRAM in collaboration with the Ministry of Science and Technology and local governments, with additional objectives to influence local societies and companies on their social responsibility for science and technology. The last four years have seen two new Science Centers in operation and one in development and another two in the initial stages of development.

Keywords: *Society and Its Culture, Science, Technology and Education, Psychographics, Competency, Resource, Agent of Development, Outreach Program, Social Responsibility.*

THE GOVERNMENT EDUCATION POLICY

Science and technology development has forced the government to have better education. To suit the demand of the market:

SCIENCE CENTER COULD BECOME A "RESOURCE"

INFORMATION TECHNOLOGY AND COMMUNICATION

Have made the society aware of what is happening in the world in a very picturous form. Science Center could become a means to relate to science and technolgy.

THE INFLUENCE OF A RAPIDLY CHANGING WORLD

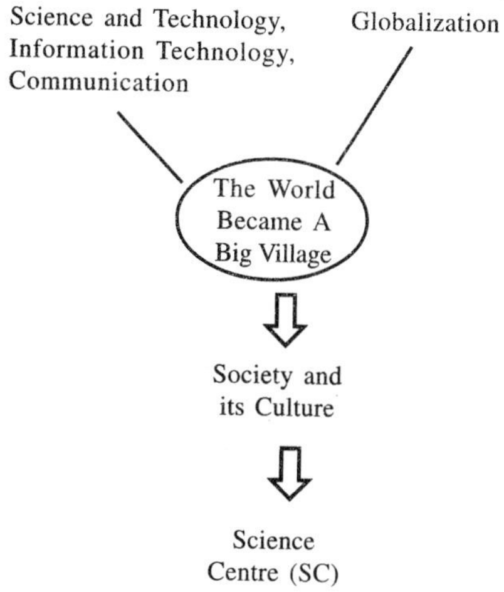

DEMAND ON LEISURE TIME

- Due to economic development, the people are able to have more leisure time.
- Knowledge on leisure time is psychographics; psychographics analysis and survey is a must for SC.

- The three variables mentioned above are important to be considered with analysis.

PSYCHOGRAPHICS: REASONS TO VISIT SCIENCE CENTER

- To look for new experiences: Interactive Exhibits
- To learn: Construction Learning
- Doing something meaningful: Family Program
- Social interaction: Science Camp
- Feeling comfortable: People Friendly Atmosphere
- To take part actively: Hands-on Activities.

SCIENCE EXHIBITS AND PROGRAMS SHOULD RELATE TO ONE OR MORE OF THE ABOVE REASONS

THE DEMAND FOR BETTER EDUCATION

The Government of Indonesia has implemented new "Competence based Curriculum". SC became a "Resource" complementary to their laboratories and learning activities, such as thematic program or demonstration, student working sheet, and creative exhibits design program for teachers.

CONTEXTUAL EXHIBITS AND PROGRAM

- Multimedia, especially television, have made the society able to follow many happenings related to their life- needs, such as environmental issues, etc.
- Contextual exhibits and programs could become a medium to bridge and explain its relation to science and technology

STRIVING FOR EQUITY

- Indonesia is striving for equity–there is no other solution than having SCs. in major cities of the country
- As a National Development Program under the Ministry of Research & Technology, Jakarta's Science and Technology Centre, PP-IPTEK, became an agent of development–through extensive outreach programs–resulted in two new SCs, one in development and two more in the initial stages.

THE IMPORTANCE OF PROMOTION

The success of SC is measured by the numbers of visitors

- Promotion division should be created independently with high position in the organization and directly responsible to the Director.
- Beside promotion to potential visitors, promotion on company social responsibility is too important to be neglected *i.e.*, sponsorship program.

7

Role of The National Science Centre in Nurturing Science and Technology Learning

Jayanath Appudurai,
Director, National Science Centre Malaysia, Pesiaran Bukit Kiara
50662 Kuala Lumpur, Malaysia
Tel. No: (603) 2092 1260, Fax No: (603) 2092 1250
E-mail: jayanath@psn.gov.my

ABSTRACT

Is there a case for science centres as non-formal learning institutions? Yes, if the raison d'etre is clearly understood by policy makers, educators, science centre personnel and the society at large. The reason for existence is none other than to make science and technology [S&T] truly relevant to the daily lives of our people. Science centres that view themselves as non-formal learning institutions formulate their vision, mission and programmes based on some implicit assumptions of the societal environment in which they operate. The National Science Centre Malaysia (NSCM) has a key role to play in contributing to the development of a science literate society. In particular, we believe that our resources should be mobilised to galvanise the potential of the younger generation of Malaysians to become science-literate citizens of the new millennium. The aim of this paper is to share NSCM's experiences in evolving a community-based learning institution and to outline the way forward for science centres in promoting and nurturing S&T learning in developing countries.

Keywords: *Science & Technology learning, Non-formal education, Life-long learning, Science-literacy, Knowledge-based economy, Human capital.*

INTRODUCTION

Is there a case for science centres as non-formal learning institutions? Yes, if the raison d'etre is clearly understood by policy makers, educators, science centre personnel and the society at large. The reason for existence is none other than to make science and technology [S&T] truly relevant to the daily lives of the public, that is, the individual visitor to the science centre. It is only when the visitor really learns to appreciate, explore and understand his or her relationship with S&T that a case can be made.

Science centres that view themselves as non-formal learning institutions formulate their vision, mission and programmes based on some implicit assumptions of the societal environment in which they operate. Some generic assumptions are:

- The public needs to appreciate and understand more about S&T facts, concepts and principles.
- The public needs to be more aware of the social, economic and ethical implications of S&T developments.
- The public's phobia of S&T needs to be addressed.
- The public needs greater access to scientific knowledge and technological skills.

Concomitantly the rationale for these assumptions is that, if these needs are addressed, then the public presumably can:

- Contribute effectively to the socio-economic development of the nation;
- Have greater confidence in dealing with the challenges of the rapidly changing technological world;
- Participate knowledgeably in discourses and debates on societal and ethical issues that arise from new developments in S&T, for example human cloning; and
- Enjoy a better quality of life.

The aim of this paper is:

- To define the context in which the National Science Centre Malaysia (NSCM) operates to promote science & technology learning among Malaysians;
- To describe the programmes and activities undertaken by the NSCM in discharging the given mandate;
- To share NSCM's experiences in evolving a community-based learning institution; and
- To outline the way forward in promoting and nurturing S&T learning in developing countries.

COUNTRY BACKGROUND

In Malaysia there are currently two science centres operating in the capital city Kuala Lumpur, namely, the National Science Centre Malaysia (NSCM) and the Petrosains Discovery Centre (PDC).

The NSCM, a government agency under the Ministry of Science, Technology and Innovation, began operations in November 1996 and is situated on a 20-acre site at Bukit Kiara **[Appendix A]**.

The PETROSAINS Discovery Centre which opened in March 1999 is owned and funded by Malaysia's national oil and gas company, PETRONAS **[Appendix B]**.

CONTEXT

The strategic vision for the nation as encapsulated in Vision 2020 is to achieve a fully developed nation status in the material, intellectual and spiritual dimensions. Underpinning this vision is the development of human resources, i.e. the capability and resilience of the individual Malaysian to operate effectively in the new millennium.

One major development thrust of the National Vision Policy (2001–2010) is to achieve a knowledge society through the development of the knowledge based economy. Central to the development of a knowledge economy is the capacity of the nation to innovate and create value in all economic sectors. Basically this means generating ideas that can be turned into innovative products and services to compete in the globalised world market.

The key to building this innovative capacity is a high level of knowledge and skills in science and technology (S&T). Without a doubt, S&T and research and development (R&D) are important determinants of knowledge and innovation generation. Thus the vision of building a science-literate society is to develop the critical mass that can engender a culture of innovation and creativity. Under the Eighth Malaysia Plan, one major focus is to increase the S&T human resources of the nation through realigning of the education system. The target is to achieve a 60 to 40 ratio of science to arts enrolment at the secondary and tertiary levels.

THE REALITY GAP

Whilst the desired end-state is clear, the reality however is that Malaysia has a long way to go in becoming a science-literate society. The research evidence thus far indicates that:

- The majority of populace seems to be unaffected by developments in S&T.
- The majority has difficulty in establishing relevance of S&T to their daily lives.
- There is a declining interest in S&T.

Interestingly, surveys on public understanding of S&T in Malaysia and in other countries consistently make two seemingly contradictory conclusions. Whilst the appreciation of S&T among the public is high, their confidence about the facts, principles and personal relevance of science lags far behind.

Admittedly, the probable causes for this state of affairs are multidimensional. However, a significant cause is the weakness in promotion and communication of S&T learning in the country.

ROLE OF NSCM

The promotion and communication of S&T learning is central to building a science-literate society. Currently several players are involved in delivering this solution. One such player is the NSCM.

The NSCM is primarily a non-formal educational institution established in 1996, with the mission to humanise the sciences. This is done through the creation of an environment of learning and discovery to enhance the level of awareness of S&T. In short, the NSCM's programmes and activities are focused on making S&T relevant to the lives of Malaysians.

In line with this, the NSCM has defined its core business as "nurturing life-long learning of S&T among Malaysians".

The NSCM's activities are basically built around the core Science Enrichment Programme and implemented through two basic strategies, namely, "In-Reach" and "Out-Reach".

As a non-formal learning institution the NSCM has adopted a three-prong learning philosophy with respect to its target customer segments. Programmes and activities are based on the following key roles:

(a) Laying the Foundation

Programmes and activities such as Little Explorer and Budding Scientist are specifically tailored for nursery and pre-school children aged two to six years. The approach is learning through play and discovery and is based on the belief that it is never too early to learn science and technology. In fact the greatest potential for planting the seeds of learning lies with this group. Importantly these activities are aimed at preparing young Malaysians for an easier passage into formal S&T learning at the primary school level.

(b) Reinforcing Learning

Programmes and activities basically tailored to complement formal learning at primary and secondary school levels. Programmes such as Science Wonders, Fun Science, and Robotics workshops are more hands-on and interactive activities meant to reinforce the theoretical learning at school. The emphasis is on creative learning processes, and not rote-learning, as there are no exams!

(c) Nurturing Life-long Learning

Programmes that are more adult, family and community-based, are primarily aimed at enhancing appreciation and understanding of everyday science as well as topics of public interest. These include talks and workshops and special programmes such as Science Fest Week and Special Citizen's programme. One important component is the professional development workshops for primary and secondary school educators.

Thematic exhibitions are also organised to cater for all customer segments. Primarily the objective is to increase awareness and understanding in specific S & T subject areas such as DNA-Genomics, Biotechnology and Aviation etc. Interactive activities that engage the visitor to explore his or her learning abilities are an integral part of all thematic exhibitions. However exhibitions such as "Saluting Malaysian Inventiveness" and "The Nobel Prize Centennial Exhibition" are expressly organised to focus on the theme of creativity and innovation.

The underlying philosophy in delivering the above programmes is that learning is a continuing and cumulative process unique to each individual. What is critical is that the programmes need to meet customers' learning expectations as users of the services of the NSC. Both the educational and exhibition programmes are aimed at providing high-value to the customer in terms of experiential learning. In short, programmes are designed to address the key components of "hands-on", "minds-on" and "hearts-on" learning experience.

EVOLVING A COMMUNITY-BASED LEARNING INSTITUTION: SOME LESSONS

Since opening its doors to the public in November 1996 the NSCM has gone through two distinct phases, namely:

Phase 1: Euphoria and Decline [1996–2000]

In 1997 NSCM as a new centre and place of attraction received 250,000 visitors. However after this initial euphoria visitor numbers continued to decline by about an average 40 percent over the period 1998-2000.

Phase 2: Revitalization and Consolidation [2001–2004]

Since 2001 visitor numbers have increased by an annual average of 20 percent over the base year 1997. In 2003 we received almost 360000 visitors and we expect to reach 500000 visitors in 2004.

Primarily the revitalization of the NSCM was a result of a strategic planning process initiated in 2001 involving all NSCM personnel. The basic thrust of the plan focused on:

(a) Formulating a vision and mission based on national science and technology policies and programmes

It was important to clearly understand and articulate our core business, that is, NSCM's reason for existence. How would NSCM contribute to the larger national agenda of achieving a science-literate society?

(b) Developing products and services based on customer needs by target market segmentation

This was critical in establishing NSCM as a community-based learning institution as we had to offer products and services that customers from different age-groups and interests really valued.

(c) Implementing a cohesive marketing, promotion and communication strategy

Positioning and branding NSCM as a place of learning for the different market segments was the key objective. Several media were used for this purpose.

(d) Establishing smart partnership networking with the public, private and non-governmental sectors

This opened up a whole host of synergistic activities that included sponsorship, collaborative programme that catered for various niche markets. Importantly this programme provided tremendous opportunities for mobilization of resources and resulted in a WIN-WIN strategy.

(e) Consolidating internal management processes

A comprehensive review and realignment of internal management processes was critical in evolving a customer-focused and market-oriented organisation. Importantly the "buy-in" by NSCM personnel resulted in higher sense of commitment and ownership.

An early indicator that these measures were achieving results was in the form of a quantum increase in requests by private sector organisations, NGOs and educational institutions to sponsor and/or organise programmes with NSCM.

Key Success Factors

In order to remain a relevant, dynamic and vibrant learning institution, our focus is on both the external and internal customers. The factors that have and would continue to determine our success are as follows:

- Relevance and timeliness of products and services to target customer segments.
- Scope and scale of out-reach programmes for rural and marginalised communities.
- Reinforcing learning and developing learning opportunities for both external and internal customers.
- Customer-focused internal organisational processes and continuing improvements.
- Enlisting commitment and resources support of key stakeholders.

Key Performance Indicators

At NSCM we have formulated indicators to monitor and evaluate our level of performance on an annual basis as follows:

Quantitative:

- Fifteen percent annual increase in number of visitors.
- Ninety percent of customers rating products and services as good to excellent.
- Increase in the scope and scale of out-reach programmes.

- Reduce cost per visitor by 10 percent annually.
- Ten percent increase in the number of institutions/organisations becoming Smart Partners

Qualitative

- Increase in the level of customer satisfaction with products and services [Weekly/monthly feedback & annual evaluation survey].
- Increase in number and value of private sector/industry sponsorship
- Increase involvement of scientists and S&T associations as collaborators.

NURTURING S&T LEARNING: SOME ISSUES

At NSCM, programmes to nurture S&T learning are based on some implicit assumptions of the societal environment, namely:

- The creative learning potential is latent in all human beings–from toddlers to senior citizens.
- The key to unlocking this potential is to engage the whole person–hearts-on, minds-on and hands-on approach.
- The specific programme content must be based on customer needs profile and segmentation.
- The key to unlocking the creative learning potential is in the mode of presentation.

Be that as it may, when it comes to the issue of engendering a culture of life-long learning, we have had to address some uncomfortable, albeit, key issues of a systemic nature:

- Does the formal system of education encourage creativity in learning?
- Do our educators view themselves as facilitator-learners?
- Do parents see themselves as having a key role as life-long educator-coaches?
- Do academics, scientists and researchers view themselves as catalysts of learning in society? and
- Does society as a whole value life-long learning? Or are rote-learning and the "paper-chase" the norm?

Whilst there are no definitive answers to these questions, suffice to say that intuitively we can all agree that these are real issues. We in Malaysia are still grappling with these issues.

WAY FORWARD

Science Centres in developing countries have a key role to play in contributing to the socio-economic development of their respective nations. In particular our resources should be mobilised to galvanise the potential of the younger generation to become science-literate citizens of the new millennium.

However in order to achieve this goal, Science Centres require the active involvement and support of several key societal players. The **scientific community**, in particular, scientists in universities and research institutions, both public and private, can play a significant role in communicating the utilitarian value of S&T to the society at large. Research results and findings can be demonstrated in a fun way to enhance the knowledge and appreciation of S&T by the public. Specifically, scientists can become role-models for young learners in order to excite and stimulate them to become scientists, technocrats and inventors of the future.

The **educators** in the formal system are our natural allies. However, in our humble opinion, the majority of educators in the formal system need to re-orient their approach to teaching-learning. The emphasis should be on facilitating learning and having the courage to say to any student: "I do not know the answer to your question but let us find out together".

The **private sector** too has a role to play in advancing the national agenda. Showcasing home-grown technology products through thematic exhibitions at the Centres can build confidence among citizens that we too can be the producers of technology, not just consumers of foreign technology.

Non-Governmental Organisations [NGOs] can contribute through specific educational activities that enhance the awareness and knowledge of the public with regard to social, environmental and ethical issues arising out of S&T developments. Surely there must be more creative ways to involve the public in dealing with the problems of environmental degradation or the fight against HIV - AIDS.

Parents, senior citizens and all members of the larger community have a crucial role as active participants in advancing S&T learning. Science Centres require this invaluable human capital to perform the role as "educators" to complement their activities. At NSCM we see tremendous potential in adopting a "smart partnership" approach to achieve the goals of a science-literate Malaysian society.

For this to become an abiding reality, all key players need to cultivate a mindset that places a premium on life-long learning. With this predisposition all members in society, be they individuals or organisations, can effectively contribute to nurturing S&T learning among citizens of developing countries.

REFERENCES

Beetlestone, J.G; Johnson, C.H; Quin, M; & White, H. (1998). The Science Centre Movement: contexts, practice, next challenges. Public Understanding of Science, 7, 5-26.

Government of Malaysia (2001) The Third Outline Perspective Plan 2001-2010, Kuala Lumpur, Percetakan Nasional Malaysia Berhad.

Malaysian Science and Technology Information Centre (MASTIC). (2003). The Public Awareness of Science and Technology Malaysia 2002, Kuala Lumpur, Dasar Cetak (M) Sdn.Bhd.

Siow Heng Loke et.al. (2003), National Science Centre Malaysia; An Evaluation of Its Role, Kuala Lumpur, University of Malaya and National Science Centre.

APPENDIX A
NATIONAL SCIENCE CENTRE, MALAYSIA

The idea of a National Science Centre was conceived in the early 1980s by the National Council for Scientific Research and Development.

The objective was to promote awareness and understanding of science and technology among Malaysians. Realizing the scale of the endeavour, a unit of government was established by the then Ministry of Science, Technology and the Environment to develop the project.

The Temporary Science Centre increased its collection of exhibits but it was not until 1990 that the National Science Centre project was approved under the Fifth Malaysian Plan. Construction on the building and basic infrastructure began in November 1991.

Work on the unique National Science Centre building was completed in 1993. A truncated cone capped by a geodesic dome remains an architectural testament to the creativity of the Public Works Department that supervised the project at every stage of the building.

Equipped with an Integrated Building Management System (IBMS), the National Science Centre has two levels of exhibition floor covering a total of 6,717 square meters. A multi-media library, a general science laboratory, three workshops, two auditoriums and a multi-purpose hall provide a range of facilities within the interior. The landscaped grounds continue the educational theme. An eco-pond, a herbal garden, and a children's playground complement exhibitions in the main building and further accentuate the emphasis on physical and biological sciences.

The National Science Centre project had a commitment to keep abreast of rapid developments within the fields of science and technology. Interactive science exhibits designed by local and foreign experts continue to stimulate, excite, and encourage visitors to have an active interest in science and technology.

Officially launched by the former Prime Minister of Malaysia Y.A.B. Tun Dr. Mahathir bin Mohamed on 29th November 1996, the exhibits are categorised into two groups. "Basic Sciences" and "Technology". The overall concept of the Centre is to link science and its applications to aspects of our daily lives.

Mission

To nurture interest in life-long learning of science & technology among Malaysians.

Objective

To increase public awareness, appreciation, interest and understanding of science and technology.

Strategies

- To provide the environment and facilities for the fun learning of science.
- To provide and fabricate interactive science exhibits.
- To provide and organise fun science programmes and activities.
- To author and publish science literature.
- To act as facilitator and adviser on non-formal science education.

APPENDIX B
PETROSAINS DISCOVERY CENTRE

What is PETROSAINS?

PETROSAINS is an interactive Science Discovery Centre that tells the story of the science and technology of the petroleum industry in a fun way. Its concept and contents revolve specifically around petroleum science, yet embrace all the general sciences.

Why was it Built?

PETROSAINS was incorporated to extend PETRONAS' commitment as a socially-responsible corporate citizen to provide a rich and stimulating environment aimed to equip Malaysians with the right knowledge and skills geared towards meeting the challenges of the new century.

At PETROSAINS we strive to inspire the spirit of innovation and leadership in petroleum science and technology. PETROSAINS aims to be a seeding platform for stimulating interest in science and technology in the hope of nurturing home-grown, world-class scientists and technologists. It is also about creating more awareness of the exciting career opportunities in the petroleum industry.

Where is it Located?

PETROSAINS is housed in the world's tallest buildings, the PETRONAS Twin Towers, in the heart of Kuala Lumpur. Entrance to PETROSAINS is through Level 4 Suria KLCC, which is accessible via various modes of public transport. The mall also boasts ample parking space for private vehicles.

Whom was it Built for?

PETROSAINS' primary target audience comprises young Malaysians who will gain from the rich source of information, presented in an innovative, hands-on learning environment.

The Centre is also built for all Malaysians, who will be able to experience the showcase of exhibits featuring the wonders of science and technology related to the petroleum industry.

Foreign visitors to PETROSAINS will be able to learn about Malaysia's rich resources and appreciate the strength, resourcefulness and pioneering spirit of its people.

How Big is it?

PETROSAINS covers a total exhibit area of 70,000 square feet. It occupies levels 4 and 5 stretching out into both Tower 1 and Tower 2 of the PETRONAS Twin Towers.

What is so special about PETROSAINS?

The Design Concept

PETROSAINS is designed as a modern contemporary museum that allows visitors to touch, feel and manipulate the items on display. Its interactive, theatrical and futuristic ambience helps visitors to be part of the museum. A visit begins with a fascinating ride and then proceeds in a linear sequence through the exhibit area filled with interactive displays. The linear design incorporates a unique storyline which provides an insight into the whole petroleum industry, from the significance of petroleum in our daily lives to the formation of petroleum and followed by petroleum exploration as well as downstream activities.

The state-of-the-art design is complemented by thematic props, enchanting backdrops, realistic visual and sound effects, three-dimensional images, simulation effects and spectacular recreations. The exhibits combine hi-tech and highly advanced presentations with simple, captivating displays and even delve into the imagery world of arts and theatre.

The Exhibits and Displays

PETROSAINS opens to its visitors an enormous window of information on the history and development of petroleum and its related products. The style and sequence of the exhibits trace the evolution of the petroleum industry and how the advancement of science and technology in the industry has tremendously improved the quality of life.

Using revolutionary methods of presenting ideas, PETROSAINS showcases its exhibits and displays in entertaining and exciting ways. For example, the theatrical setting of "Wayang Kulit", or the traditional Malay Shadow Play, is used to demonstrate the uses of petroleum and educate visitors on the significant role petroleum plays in everyday life.

A special area is the Geotime Diorama which is packed with realistic models of prehistoric animals and plants and a volcano. A feature that would delight visitors–especially the young ones–at the Diorama is the "Singing T-REX".

Another interesting replication is the "true-to-life" oil platform. Visitors will get the rare opportunity of being on an oil rig, see every aspect of what life is like on it as well as experience a simulated helicopter ride to the oil platform.

Other unique features include the Water Sculpture in the foyer which gives an "under the sea" effect to visitors upon entering PETROSAINS. After all, most of Malaysia's petroleum resources come from under the beds of the South China Sea.

The PETROSAINS People

The unique and interesting exhibits and displays at PETROSAINS are made more fun and exciting by its team of dedicated front-line personnel. They act as guides, presenters and demonstrators as well as facilitators to ensure that visitors' enjoyment and learning experience are maximised.

Internal & Outreach Programmes

To encourage and nurture interest in science and technology, particularly in the petroleum industry, PETROSAINS has developed various internal and community outreach programmes designed for school children as well as students of higher institutions.

How to Get in Touch with Us?

Address: PETROSAINS Sdn Bhd, PETRONAS Twin Towers, Level 4,
Suria KLCC, 50088 Kuala Lumpur, Malaysia

Fax: (603) 2331 2903

Infoline/Booking: (603) 2331 8181

SUMMARY

PETROSAINS offers visitors a truly new, innovative way of learning about the petroleum industry and the wonders of science and technology. With its interactive educational displays and entertaining showcases, PETROSAINS aims to contribute to the development of a society knowledgeable in science and technology and complement Malaysia's efforts towards realising its Vision 2020.

8

Current Status of Science Centres and Museums in Mauritius

Oree Vishwamitra
Curator, Rajiv Gandhi Science Centre, Old Moka Road, Bell Village, Mauritius
Tel: (230) 2132773, Fax: (230) 2132726, E-mail: rgsctf@intnet.mu

ABSTRACT

The experience of countries that have succeeded in attaining a high level of sustained economic development underlines the crucial role of science and technology in that success. Emerging economies have long understood that science education is important for generating a scientifically literate workforce, capable of adapting to medium and high technology sectors. On the other hand, science centres and museums have a vital role to play in the field of science education. They are recognised as centres of learning for the child in that they inspire him/her to discover his/her own creative potential. For the teacher too, they have become resource centres on which he leans heavily for acquiring knowledge and experience as well as innovative and creative methods of science teaching. As such, science centres and museums certainly have their share of responsibility in providing basic scientific literacy to the population at large in Mauritius, a small island of only 2040 square kilometres, which aspires to joining the bandwagon of technology-driven states. In this report, the circumstances that have led to the setting up of the first science centre in Mauritius are detailed. Then, a comprehensive description of the science centre including its objectives, galleries, activities and facilities is given. Next, an account of other museums, with the same endeavour to popularise science and technology among the general public is presented. In conclusion, the approach and outlook of the science centre and other museums in Mauritius as the country sets on a critical phase of its development are described.

Keywords: *Mauritius, Rajiv Gandhi Science Centre, Museum, Technology, Exhibits, Education.*

INTRODUCTION

In today's global economy where competition is becoming fierce and where development and diffusion of new technologies are central to growth of output and productivity, it is generally agreed that the survival of small island states like Mauritius will depend considerably on its capacity to become technologically innovative in all major sectors of the economy [Mauritius Research Council (MRC), 2000]. The Mauritian Government has been quick in appreciating this fact and has initiated a paradigm shift in its economy: low and medium technology manufacturing are making way for high technology industries such as information and communication technologies and biotechnology. Bringing forward this agenda calls for a good public understanding of science and technology because it is bound to raise some uncomfortable issues. Mauritius is on the brink of exciting developments in science that will affect everyone's life and along with the benefits will come potential risks which would cause understandable concern. People rightly expect proper safeguards against risks, and full and open discussion about the values, benefits and risks of scientific developments. Moreover, only a scientifically literate workforce will be able to adjust to the high technology industrial sector where constant upgrading of skills and knowledge is essential (Seegobin, 2004).

With a view to assessing the status of science and technology in Mauritius in general, and more specifically in the educational sector, the Mauritius Research Council, in its capacity as the adviser to Government on matters pertaining to science and technology, undertook a number of studies. The findings of theses studies were alarming as it showed clearly that not only science is becoming less attractive in schools but public understanding of science is as well on the decline. For example, fewer than thirty percent of Ordinary Level students opt for science subjects in the secondary schools.

Among the reasons identified for the lack of interest in science subjects were the dearth of job opportunities and the perception that science subjects are difficult. In fact, such a trend is reflected worldwide. In the wake of these problems, the Rajiv Gandhi Science Centre (RGSC), the first science centre in Mauritius, has been set up with the objective of promoting science and technology in Mauritius in a non-formal way.

RAJIV GANDHI SCIENCE CENTRE

A relatively new concept in developing countries, science centres are fast becoming model institutions for non-formal learning alongside their more traditional role of being custodians of the nation's scientific and technological heritage. While the traditional science museums have the emphasis on the preservation of objects of scientific value, science centres stress on the exploration of the logic of science through the interactive participation of visitors. A sense of exploration and discovery is in-built in the approach of a science centre rendering the learning process self-motivated and pleasurable. Possibly, no other institution imparts science experience so effectively for the community as a science centre. Ultimately, the aim of a science centre is to act as a cradle for creative and independent thinking conducive to decision-making by the people in respect of the various scientific projects in their country.

Figure 8.1: Front View of RGSC Building

Objectives

When a proposal for setting up a science centre in Mauritius was first mooted, the Indian government readily agreed to assist in implementing the project and identified the National Council of Science Museums (NCSM), Kolkata, as the nodal agency in India for rendering such assistance to Mauritius. The NCSM subsequently submitted a project proposal which was duly considered by the Mauritian Government. This paved the way for the implementation of the first science centre in Mauritius. In recognition of the deep involvement of India in the project, the science centre was named Rajiv Gandhi Science Centre after the former Prime Minister of India.

The main objectives of the centre are to:

- Promote science and technology through various programmes, activities and exhibitions,
- Inculcate awareness in scientific fields among the Mauritian public in general,
- Supplement school education in a non-formal way by programming science demonstration lectures, science fairs, science seminars and other outreach programmes for school children,
- Inculcate creativity among young people, and
- Popularise science with the aim of enhancing public understanding of science and technology (RGSCTF Act, 2004).

In order to achieve these goals, the RGSC relies on its innovative, visitor-actuated exhibits together with its interactive educational programmes.

Galleries

During the conception stage of the science centre, one of the challenges facing the stakeholders was to maintain a reasonable number of visitors annually. The population of the country being about 1.2 million, there was a fear that the visitor figure will drop after a few years of operation. Yet, an evident source of potential visitors would be the growing number of tourists visiting the country every year. In 2002, for example, nearly 700000 tourists were reported to have been in Mauritius.

But, then, how to urge the tourists to leave the sea, sun and sand for a visit to the RGSC? An ingenious solution was to offer tourists the opportunity to discover Mauritius under one roof in the science centre itself. In this respect, the three galleries on the ground floor provide the visitors with an enlightening and entertaining corridor of experience-based knowledge on various facets of Mauritius. Two other galleries on the first floor introduce the public to the emerging technologies that are affecting our way of living and to a playground of hands-on science where the play will kindle their spirit of enquiry.

Figure 8.2: Students Learning About the Formation of the Solar System

Origin of Mauritius

The Origin of Mauritius gallery invites the public to a journey of exploration in a quest to unravel the mysteries of astronomy. Starting with the opening seconds and minutes of time which saw the formation of the first atomic particles, the journey proceeds with the various stages through which the Universe went in about 14 billion years of its existence. Having understood where the Universe came from, the visitor can contemplate the various theories regarding the Universe's ultimate fate. In the immensity of the Universe, the visitor learns about the countless galaxies of which one, the Milky Way, is of specific interest to us. The visitor can witness the nine planets of the solar system rotating around the Sun in a 3D-model.

An interactive software "Our Address" takes people on an exciting tour from the edge of the Universe (10^{25} m) to the atom (10^{-16} m) in successive steps of 10 metres, giving them a notion of distances on an astronomic scale. The gallery also contains an instructive walkway of exhibits which highlights diverse features which have marked the history of the Earth. While "Wandering Continents" explains chronologically how the continents have been grouping and then moving away from each other to change the face of our planet throughout its lifetime, "Evolution of Life" describes the evolution of life on Earth.

The gallery culminates in a video projection simulating the volcanic eruption which gave birth to the island of Mauritius. To sum up everything, a multimedia show "Story of the Universe" presents a stimulating mix of lively video projections, colourful slides and expert, jargon-free voice-over to review the important concepts introduced in this gallery.

Land and Environment of Mauritius

Mauritius is a small island of volcanic origin in the sub tropical region. Over the ages, unique fauna and flora have evolved with no interference from man, nature being the sole agent of change of the environment. However, the subsequent arrival of man and the establishment of permanent human settlements have had drastic effect on the environment. In this context, the Land and Environment gallery endeavours not only to inform the public about their environment but also to galvanise them to

Figure 8.3: A Member of the RGSC Staff Initiating Visitors to "Demography" Exhibit

make earnest efforts to protect, preserve and enhance the quality of the environment.

A 3-D map underlines the geological features of Mauritius including the highest peak, volcanoes, caldera rims and trachyte domes. The effect of weathering on the landscape is clearly illustrated through a double-diorama. Lava tunnels, which are subterranean passages created from hot molten lava flow thousands of years ago, are considered among the most prominent geological features of the island. In this gallery, visitors can venture into a mock-up lava tunnel to unravel its secrets. After a tour inside the lava tunnel, the public can appreciate samples of old, intermediate and recent volcanic rocks.

Large aquariums, projection areas, coral specimens and an interactive software further delight the public as they discover the marine ecosystem of Mauritius. The next set of exhibits aims to sensitize the public about the need to take care of the environment. A replica of a dodo is strategically situated, standing as a sentinel in the gallery to epitomize extinction due to irrational exploitation of fauna and flora by humankind. "Protecting the environment" explains how the indiscriminate use of Earth's resources has resulted in undesirable changes in natural parameters. The numerous consequences of these changes are portrayed in "Price of Progress": greenhouse effect, ozone layer depletion, acid rain and extinction of species, just to name a few. A fully automatic weather station gives visitors real-time information on the weather. Some historical paintings of the landscape of the island and statistics about the demography, education and health systems get the visitors acquainted with Mauritius.

Before leaving the gallery, people can test their knowledge about the land and environment of Mauritius in the "Quiz Corner". Here, multiple-choice questions are projected onto a giant screen and each team can enter its answer via a console. The smooth running of each quiz session is left to a computer programme, which additionally provides full multimedia support.

Resources of Mauritius

Traditionally, the Mauritian economy was based on sugar production and export but, over the last 20 years, it has diversified. The Resources of Mauritius gallery guides the visitor through the whole gamut of resources which has sustained significant economic growth in the island. At the entrance, an interactive software providing a comprehensive overview of all the sectors of the economy, supplements the knowledge imparted by the exhibits.

Mauritius being an island, marine resources are bound to grab the attention of the public. To this effect, translides, documentary films and visitor-operated displays portraying the rich marine biodiversity of Mauritius form an integral part of the gallery. Obviously, sugar takes the centre stage in the gallery. A scale model of a sugar factory elucidates the stages involved in sugar production. Many of the special sugars that are produced locally are displayed.

Two large multimedia projection areas highlight the significant impact that tourism and the multicultural Mauritian society have on the economy. In the mid-eighties, the textile sector was the key factor in steering the country to an economic boom. The RGSC naturally devotes a cluster of panel-based exhibits enhanced with thematic films to help visitors get more insight into the textile industry. Finally, in appreciation of the vital contribution of the Indian government to the RGSC project, a software presents thorough information on other projects that have been implemented in Mauritius as a result of Indo-Mauritian collaboration.

Fun Science Gallery

The Fun Science gallery incites the visitor to embark on an enlightening tour where science is exploration, discovery and fun. People are encouraged to handle the forty-one participatory exhibits and to try out all the options in an attempt to see the miracles of science come alive before their eyes.

A first series of exhibits tries to unravel the mysteries of sound. Anyone can compose his own music by striking freely hanging metal pipes of different lengths with a wooden hammer, observe his voice being split into its constituent frequencies as he speaks in a microphone or investigate the working principle of loudspeakers by observing a continuous flow of music coming out from a plastic bucket placed on a vibrating pin. While magnetism usually dwells unseen in the atoms of materials, in the Fun Science gallery, a set of exhibits provides a fascinating experience to the visitor as he observes magnetic field lines run from a magnet's north pole to its south pole, guesses the shape of a magnet from its magnetic field pattern, familiarises himself with the principle of operation of the three-phase induction motor and explores funny applications of eddy currents.

The properties of light have puzzled man since time immemorial. One corner of this gallery unveils some of these properties. People can combine red, green and blue to create a whole array of colours. Optical illusions keep many visitors pondering for a long time. For instance, a human head being offered on a platter for lunch or curved mirrors that reflect distorted images are mind-boggling. The Fun Science gallery abounds with other exhibits through which discovering the complex rules of probability, learning about the properties of an ellipse, inquiring into the principles underlying flotation, determining why planets remain in orbit around the Sun, understanding the process of copper plating and finding out why some liquids make an impossible mixture turn out to be exciting games of fun.

Frontiers of Modern Technology

The "Frontiers of Modern Technology" gallery offers an initiation to some of the breathtaking technological frontiers at the edge of our knowledge envelope. As the visitor walks into the gallery, he immediately confronts the foremost concern for our quality of life: the rapidly escalating world population.

The first group of exhibits gives him a clear understanding of the means and ways in which the key resources are optimised to provide for the basic needs of this increasing population. Thus, after "Feeding our World of Teeming Millions" has reviewed the evolution of the world population back from the dawn of mankind to present day, "Harvesters of the High Seas", appreciate how ships use state-of-the-art technology to precisely locate shoals of fishes in high seas, "Dairy Farming" investigates the application of new technologies to increase the milk output and "Technologies that changed the Face of the World" portrays the engineering marvels represented by skyscrapers, bridges, roads, canals, pipelines, dams and tunnels. In the latter context, "The Changing Face of Port Louis" reveals the stunning change undergone by the Mauritian capital through the last two centuries by means of a double-diorama and vocal explanations.

"The Textile Revolution", through real life-size models and translides, rolls the tape back in time to discover the major machines that have marked the textile industry: from the spinning jenny of James Hargreaves to the contemporary computerised spinning and weaving machines. The magic of two medical devices that enable doctors to see within the human body without surgery is revealed in "Seeing the Unseen". Here, visitors are motivated to participate in the live demonstrations with the ultrasonography and the fibrescope to understand how ultra sound and light respectively are used to reveal inner organs of the body. Still in the medical domain, "Technology of Life" provides comprehensive information about the DNA, this fundamental chemical unit of life. A giant kinetic structure of the DNA immediately kindles all inquisitive minds and drives them to learn more about cells, genes and proteins.

The next section of the gallery introduces two fields that have been on the forefront of scientific news during the latter half of last century: space exploration and nuclear energy. Space exploration begins with a trip through the history of spaceflight, starting from the 1000 B.C when the Chinese are believed to have used gunpowder rockets to the 21st century when the prospect of manned flight to Mars is looking increasingly probable. Visitors can find out how rockets are able to break the grip of Earth's gravity and send spacecrafts in outer space. One can admire a scale model of the gigantic Saturn V that launched the Apollo astronauts to the moon. The exhibit "Servants in the Sky" illustrates the growing influence satellites have on human life by providing weather and environment information, fast communication links, accurate astronomical observations and military data. One of the major requirements for sustaining human progress is an adequate source of energy. An alternative that has been the cause of much controversy during the past century is nuclear energy. A group of exhibits explains the harnessing of nuclear energy. The last section of the gallery is in effect, a corridor into the dazzling world of electronics where fun is merged with discovery and learning. The use of information technology in areas as diverse as communications, electronics, health care, science, entertainment, government, and electronic commerce is illustrated via extensive software, touch-screen computers and experience-based exhibits. "Virtual Reality" is undoubtedly the most successful exhibit among people of all ages as it demonstrates the future of the interface. The visitor can enjoy a breathtaking virtual game of football, beach-volley ball or basketball.

Science Park

The Science Park is an open air laboratory which promotes discovery by free play. The element of play develops motivation, enthusiasm, imagination and divergent thinking in a total environment where landscaping, plantation and water bodies merge with the exhibits. Since the dissemination of information is taking place in an informal and picturesque setting, understanding of science is not only easy but also long lasting. The exhibits spur adults and children alike to learn the principles of mechanics, sound, light and vibration among others.

Swings of varying lengths urge the visitors to note the basic laws of the pendulum. Two persons can experience sympathetic oscillations by sitting on two swings hanging from a common bar. Children can impress their friends by moving weights using pulleys and inclined planes. Such experiences give the child enthusiasm to experiment with other things. Exhibits such as "Sun Dial", "World Clock" and "Mobius Band" nurture a scientific attitude and a spirit of enquiry among visitors.

The properties of light are investigated in a series of exhibits. Visitors can watch the panorama through filters of different colours, rotate a plate with a bird and a cage painted on its two sides to get the impression that the bird is inside the cage or see an inverted image of the landscape in a large camera. Optical illusions are further illustrated in "Perception of depth" and "Straight line or Circle". The visitor can also try the Archimedes' screw, named after the famous Greek philosopher who devised that ingenious method of drawing water from a deep well.

Outreach Programmes

The prime reason for introducing outreach programmes is to disseminate the basic tenets of science and technology to a larger segment of the nation. Spearheading the outreach activities of the RGSC, the Mobile Science Exhibition (MSE) Unit has 24 exhibits mounted on a specially designed bus.

To foster effective two-way communication with the participant, MSE exhibits demand specific consideration in their design and presentation. In particular, they are simple, easy to handle and take an active role in literally conversing with the participant. The current set of exhibits illustrates the workings of the different organs inside the human body. The scientific facts causing apparent physical differences among respective individuals are highlighted. The message conveyed is that regardless of race, caste or colour, the organs in all human bodies perform the same functions and our external differences are due to our genes. The MSE Unit also carries a portable planetarium and basically targets lower secondary students in order to arouse an interest in science in them.

Educational Activities

The education system in Mauritius by and large has been found to be rigid and examination oriented, giving little or no scope for nurturing creativity among children.

Yet, there is a natural instinct in the young ones to explore newer vistas to satisfy their curiosity. When an element of play is coupled to this exploration, it allows children to be more creative and helps in the development of their linguistic, cognitive and social skills. Considering all such aspects, activities mainly aimed at developing creative faculties in children have been regularly chalked out by the RGSC.

Figure 8.4: "Bird in a Cage" Illustrates an Optical Illusion

Figure 8.5: Students Interacting with the Exhibits of the MSE Unit

Figure 8.6: A Planetarium Session Conducted by Trained RGSC Staff

During school holidays, the Vacation Workshop is a regular feature which attracts lots of students. They are exposed to lectures, model-making sessions and science games.

Another programme is the Science Demonstration Lecture involving two components: the software consists of a well composed script for the lecture, while the hardware is a set of prototype teaching aids developed to demonstrate a particular topic of the school syllabus. The emphasis is on a high-drama presentation, involving the students' participation and aimed at having a lasting impression on the young minds. The teaching/learning process entails a high level of activity, both intellectual and emotional, so that the overall presentation resembles a magic show rather than a classroom lecture.

The Popular Science Lecture (PSL) is another prevalent activity of the centre. It is basically a discussion conducted by subject experts on a particular science topic. PSLs are normally organised on the occasion of some major scientific event to widen the horizon of the audience. Information supported with demonstrations on themes of common interest are found extremely popular among students and general public. Finally, sky observation programmes are scheduled on a regular basis for the public to gaze at the moon and planets of the solar system through the telescopes of the centre.

As the RGSC matures, it is expected to elaborate more experiment-oriented activities which will not only develop scientific awareness and creativity in the child but also leave a sharp and distinctive mark in his entire way of problem handling at various levels.

Figure 8.7: Participants Playing with Bubbles during a Vacation Workshop

Figure 8.8: The General Public Observing the Transit of Venus by Projection Method

Other facilities

One of the most delightful facilities available in the RGSC is the Activity Area, a platform for constructive play which contributes to the cognitive development of the child. It provides opportunities to perform experiments and to undertake projects of investigative and creative nature. The Temporary Exhibition Area accommodates thematic exhibitions regularly in order to mark special scientific events while simultaneously encouraging visitors to come to the centre again. Other facilities available include the cafeteria, souvenir shop, library and auditorium.

The RGSC intends to network with like-minded organisations to bring rich, informative and varied experiences to students and the general public.

OTHER MUSEUMS IN MAURITIUS

Museums are integral parts of the Mauritian society. The country has such a rich cultural heritage that museums honouring it tend to attract much attention. Yet, there are in the country several museums dedicated to furthering public understanding of science in its various forms. The more traditional role of collecting, preserving and interpreting technical artefacts has been abdicated, to a large extent, to historic museums, such as the Natural History Museum and the Postal Museum, both located in Port Louis. Moreover, excellent zoological parks, a priceless botanical garden and some nature preserves depict other aspects of natural sciences that merit notice. The recent years have also heralded the setting up of museums highlighting the principal industries of the country. The avowed purpose is to foster among the people, knowledge of the technology applied in these sectors. A brief description of these museums is given in what follows.

Historical Museums

The Natural History Museum

The Natural History Museum, set in the *Jardin de la Companie* in the centre of Port

Louis, is the oldest museum in Mauritius and was opened in 1842. Its collection dates from early 1800s and contains unique and rare specimens of the fauna of Mauritius and surrounding islets. It gives great attention to fossils and the stuffed replica of the dodo is certainly its main attraction. This big, flightless bird disappeared shortly after human settlement in the island at the end of the 17th century.

The museum also houses stuffed representations of several other extinct birds as well as specimens of animals and fish that are still with us. The Natural History Museum has a large attendance of tourists, in addition to the numerous school visits. It operates under the purview of the Mauritius Museum Council, a corporate body created by the government under the Ministry of Arts and Culture to manage public service museums in the country. The vision of the **Council** is to be a centre of excellence in the quest for and dissemination of knowledge on the collective memory and in the preservation of the cultural and natural heritages of the people of the country. It also runs a library with a collection of over 50 thousand books including a special section for researchers of the Mascarene archipelago (Tropicscope).

Postal Museums

Mauritius claims an illustrious history in philately. Besides being the fourth country in the world to issue postage stamps along with the United States, it was also the first British colony to do so. In this regard, the setting up of a Postal Museum was justified. Its main goals are to reveal the postal and philatelic heritage, to stimulate interest in philately and to promote the collection and study of Mauritian stamps and philately. Among the items on permanent display, the public can find philatelic archive paraphernalia dating back to the mid-nineteenth century, letter boxes, telegraph machines, stamp vending machines, stamps, postal stationery, original artworks and printing plates. Philatelists can also buy materials to mount or complete their collections.

The highly touristic Caudan Waterfront in Port Louis features the Blue Penny Museum, which is primarily devoted to Mauritian art and history, but whose place of pride is given to two of the rarest stamps in the world: the Mauritian 1847 "Post Office Penny Orange" and "Twopence Blue", estimated to be worth more than US$1million apiece. The museum is run by the Mauritius Commercial Bank, which formed a consortium of local companies to purchase the stamps at an auction in Switzerland in 1993, and bring them home for posterity.

Science and Technology Museums

Sugar World Museum

The Sugar World museum brings visitors to the fascinating world of sugar in the former Beau Plan sugar factory. As mentioned earlier, the history of sugar cane is closely linked to the history of Mauritius because it has been, for long, the backbone of its economy. The main objective in establishing the museum was to create a cultural location where the public could learn about both the historical and contemporary role of the sugar industry in a playful and relaxed environment. The choice of an old sugar factory as a site for hosting the museum also contributes to generate a spectacular technological impression.

The technological tour occurs in two phases: on the one hand, "from the cane to the juice" illustrates the transformation of the cane to the liquid state in mills, and on the other hand "from juice to crystals" portrays the processes entailed in the change from the liquid to the solid form. Attractive exhibits, systematically arranged to illustrate the evolution of technology in sugar production and supported by explanatory graphics, pay a tribute to the ingenuity of Mauritians through the ages. Instructive exhibitions reveal the applications of by-products of sugar production in various sectors. The impact of sugar on the historical, economic and commercial outlook of the country is also exposed. In a section reserved for children, a cartoon traces the different stages through which sugar goes before landing in their cup of tea. The ambitious Sugar World Museum project was realised by a private group consisting of several local stakeholders of the sugar industry.

Figure 8.9: Front View of the Sugar World Museum

Figure 8.10: Exhibits Stimulating the Inquisitive Minds of Children at the Textile Museum

Figure 8.11: A Stunning View of the Giant Water Lilies in the Pamplemousses Gardens

Textile Museum

The textile industry is one of the pillars of the Mauritian economy, enabling nearly a quarter of the population to earn a living. The textile Museum, through multimedia support, takes the visitor on a journey through time to see the evolution of fabrics, colour and texture. The various stages leading to the manufacture of garments, from the early days to the latest fabrics are detailed. The visitors can witness the dexterity of the Mauritian knitters as they manufacture pullovers in situ. A projection room equipped with a giant screen unveils the industrial world of textile manufacturing from the arrival of raw materials to the export of finished garments.

Zoological Parks and Gardens

Sir Seewoosagur Ramgoolam Botanical Gardens

The Royal Botanical Gardens of Pamplemousses, ranking among the finest botanical gardens in the world, were established in 1735 by *Mahè de Labourdonnais* around his residence. Initially a vegetable garden, the grounds were gussied up by French horticulturalist Pierre Poivre in 1768, and he laid the seeds for its present success as a garden of international acclaim. In 1849, a British horticulturalist, James Duncan, took over. The two horticulturalists introduced plants from all over the world and raised indigenous species, the fruits of which can be admired today. They include 80 palms and about 25 species indigenous to the Mascarene Islands, amongst them stately palms, fruit and spice trees, ebony, mahogony, latania and pandanus. The gardens occupy some 60 acres of beautiful landscape. One of the most impressive sights is the peaceful pond concealed by the enormous floating leaves of the Giant Amazon water lily. The flowers open white but fade into a dusky pink by the end of the second day.

The collection of palms is extensive, amongst them being Royal Palm, Queen Palm from Brazil, Raffia Palms from Madagascar, Lady Palm from China and the Talipot Palm which dies after giving a giant flower when it is between 40 and 60 years old. Visitors can also catch glimpses of Mauritian wildlife in the enclosures of the Java deers and giant tortoises. The famous gardens were renamed Sir Seewoosagur Ramgoolam Botanical Gardens in 1988 in honour of the first Prime Minister of the country (Encyclopaedia Mauritiania, 2002).

The Casela Bird park

The Casela Bird park, set in the western Black River district, is a tribute to nature. 90 aviaries hosting around 1500 birds, a variety of mammals and a petting farm beautifully set amongst abundant tropical flora constitute an enthralling experience for the public. One of the rarest birds in the world, the *Mauritian Pink Pigeon*, can be admired there. Casela also houses other animals such as tigers, monkeys, lemurs, deers and giant tortoises. Its natural setting, spacious layout and luxuriant vegetation help to make it a popular spot among the local population and tourists. It should also be pointed out that there exist numerous bird watching sites in Mauritius where nature lovers can view rare species of birds breeding in the wild.

The "Domaine du Chasseur"

In the south eastern part of the country, it lies high in the hills. It consists of 3000 acres of tropical forest inhabited by hundreds of Javanese stags, wild boars, monkeys and thousands of birds.

La Vanille Crocodile and Tortoise Park

La Vanille Crocodile and Tortoise Park is established in a rainforest valley in the south of the country. The park breeds Nile crocodiles and Aldabra tortoises. La Vanille is today recognised as the world's largest captive breeding centre for Aldabra tortoises. Other species of animals can also be found: bats, mongooses, monkeys, deers, wild boars as well as domestic animals such as goats, donkeys and sheep. An exhibition room features tanks with most of the fresh water fish, crustacean and eel species found in Mauritius. A collection of skulls from whales that have washed ashore in Mauritius over the years can also be admired. An insectarium and butterfly farm also form part of the park.

The Mauritius Aquarium

The Mauritius Aquarium at Pointe aux Piments, has adopted a new ecological system of operation. Numerous tanks and pools housed in five buildings contain examples of the rich aquatic life found on and around the tropical island. Visitors can view a variety of deep sea creatures in the largest tank. A "touch pool" gives the younger ones an opportunity to interact with various species of fish.

CONCLUSION

As indicated in this report, the RGSC and other science and technology museums in Mauritius are making great progress in their efforts to improve and expand their offerings. At the RGSC, indoor and outdoor exhibits as well as educational programmes are so designed that they motivate the general public to explore and discover by exploiting all their sensory faculties. Other museums, both long-standing and recent ones, have also adopted the participatory approach. Consequently, instead of the public looking at science and technology developments as a distant spectacle beyond their understanding, they are made to absorb the impacts of science and technology as part of their life processes. Regardless of the scope or method, the overall goal is to do a better job of furthering public understanding of science and technology in order to equip the country with adequate resources as it tackles a new orientation in its economy.

REFERENCES

MRC, 2000. Science, technology and innovation survey, Volume III. Mauritius Research Council.

MRC, 2004. Teaching and learning of science in schools in Mauritius, Volume I, Recommendations and Action Plan

Seegobin, R., 2004. Cost effectiveness study on upgrading of science laboratories in secondary schools of Mauritius, Final Report, Mauritius Research Council

RGSCTF Act, 2004. Rajiv Gandhi Science Centre Trust Fund Act 6 of 1994. Revised Laws of Mauritius 2000 Volume 6.

Encyclopaedia Mauritiania, 2002 Pamplemousses Botanical Gardens

http://www.encyclopedia.mu/Nature/Geography/Places/Pamplemousse.html (accessed 04 Oct. 2004)

Tropicscope. Mauritius places of interest–museums, Natural History Museum

http://www.tropicscope.com/eventreadmore_clients.asp?cID=386&ID=711 (accessed 30 Sept. 2004)

Abbreviations

RGSC	:	Rajiv Gandhi Science Centre
NCSM	:	National Council of Science Museums
m	:	metre
3-D	:	Three-dimensional
DNA	:	Deoxyribonucleic Acid
PSL	:	Popular Science Lecture
US$	:	United States Dollar

9

Science Centres and Museums in Nepal

Dayananda Bajracharya
Royal Nepal Academy of Science and Technology,
Khumaltar, Satdobato, P.O. Box 3323, Kathmandu, Nepal
Email: dbajra@mail.com.np, ronast@mos.com.np

ABSTRACT

Since the establishment of the first museum, The National Museum, in 1928, several other museums have been established in Nepal, most of them in Kathmandu, the capital city. Most of these museums exhibit objects related to art, architecture, culture, religion and history of the country. Some of them highlight the life of past kings and contain their personal belongings. Only very few of these museums contain exhibits related to science. Such science-related exhibits are mostly limited to plants, animals, minerals and metal works. Some science-related collections are also stored in departmental museums of educational institutions. In spite of ongoing efforts for more than two decades, Nepal still lacks a true science centre or museum. In order to fulfil the need for such a centre, the Royal Nepal Academy of Science and Technology is making a serious effort to establish in its premises a Science Learning Centre with the objective of enhancing public understanding of principles and benefits of science and technology. The present paper will provide information on some museums in Nepal that are involved in science-related activities and the efforts that are being made to establish science centres and museums in Nepal.

Keywords: *RONAST, Science Learning Centre, Science Popularization.*

INTRODUCTION

Bordering with China in the north and India in the south, east and west, Nepal is a country of amazing extremes. Its elevation ranges from 70 ms above sea level in the south to the highest point on earth in the north, the Mount Everest, at 8,848 ms, within a distance of 150 kms. This extreme variation in altitude has resulted in climatic conditions ranging from sub-tropical to arctic. Nepal has the greatest mountain ranges on earth. Its scenic beauty with forested hills, snow-capped mountains, frozen valleys, thundering rivers, rich flora and fauna, is world-famous and has few parallels elsewhere.

Within this spectacular geography is also one of the richest cultural landscapes anywhere. It is a great repository of art and architecture and also the cradle of two major religions of the world:Buddhism and Hinduism. The blending of these two great religions has resulted in a unique cultural legacy and a great array of arts and artefacts. Nepal is among the very few countries in the world, which has seven World Heritage Sites within a radius of 20 kms. The valley of Kathmandu, the capital city, where an extraordinary civilization flourished for more than two thousand years, is like an open museum with innumerable pieces of artworks scattered all over. It is said that "there are as many temples as houses and as many gods as people" in Kathmandu. With all the beautiful landscape, culture and outdoor temples, many people find the area around the museums more entertaining and educative than the museums themselves.

MUSEUMS IN NEPAL

The National Museum is the oldest, largest and the only multipurpose museum of the country. It was established in 1928 as an arsenal museum. It was initially known as the "*Chhauni Shilkhana*", which literally means "the store house of arms and ammunition". It was opened to the public in 1938 and named "*Rastriya Shangralaya*" or The National Museum of Nepal in 1967. The museum is considered the most important museum of the country. It has a large collection of weapons, including brass canons and automatic electric machine guns made locally. The museum has also on display one of the first automobiles, a symbol of modern technology, that was brought to Kathmandu in 1916 by porters carrying it on their shoulders since there was no road access to Kathmandu at that time. It also has a large collection of art and antiquities of historical, cultural and religious importance. The museum is a unique treasure house of medieval and modern works in wood, metal/bronze, stone work and paintings. There are some plant specimens and stuffed animals representing the flora and fauna of Nepal. Philatelic items are the new additions.

Other important museums that contain objects related to art, architecture, culture, religion, history etc. of the country are Patan Museum, National Art Gallery, Pujarimath Museum, Bronze and Brass Musuem, National Bronze Art Museum, National Woodcarving Museum, National Numismatic Museum, National Library, Kaiser Library, Asa Archives and National Birendra Art Gallery. There are few museums like Tribhuvan Memorial Museum, Mahendra Memorial Museum and Birendra Memorial Museum that highlight the life of past kings and contain their personal belongings. All these museums are located in the Kathmandu valley. Dhankuta Museum, Hattisar Museum, Elephant Musuem, Mustang Eco Museum, Tharu Cultural Museum, Pokhara Museum and International Mountain Museum, Pokhara, are some of the museums outside the Kathmandu Valley.

SCIENCE CENTRES AND MUSEUMS IN NEPAL

The main task of science centres and museums is to communicate a scientific spirit and mentality, to arouse natural inclination for science, to give latest information on the progress in science and technological developments and to make people understand, appreciate and conserve nature. In spite of ongoing efforts for more than two decades, Nepal still lacks a true science centre or museum that fulfils the above functions. Nevertheless, there are some museums in Nepal that contain collections related to science and technology. The National Museum has some collections related to flora and fauna of Nepal. Its rich collection of firearms provides some information on how technology of weaponry developed over the centuries. The National Bronze and Brass Museum provides information on development of metallurgy in Nepal.

The International Mountain Museum in Pokhara was opened in 2004 with the objective of representing mountain geography, ecology, people and culture. The museum also includes exhibits related to the origin, geo-science, flora and fauna of mountains, particularly Nepal Himalayas. The museum that can perhaps be truly called a science museum is The Natural History Museum. It was established in 1975 under Tribhuvan University. Nearly 60 per cent of the recorded biodiversity of Nepal is represented in the museum. It has a total collection of nearly 55 000 specimens, including 40 000 zoological specimens, 7 000 botanical specimens and 500 fossils. The museum also houses some very rare fossil specimens such as the eleven million years old tooth of the prehistoric man called *Sivapithecus* discovered from Central Nepal. The Pokhara Museum in the Prithvi Campus of Tribhuvan University has the best collection of butterflies and moths found in Nepal. The Forestry Museum of Hetauda Campus of Tribhuvan University has collections related to forestry, including wildlife. Similarly some science-related collections are also stored in departmental museums of other educational institutions. Some government departments have also some science-related collections. The National Herbarium at Godavari is a repository of plant resources of Nepal. Established in 1961, it has a collection of nearly 100 000 plant specimens collected from all over the country. The Department of Geology and Mines has a rich collection of mineral resources of the country.

The need for a science museum was felt a long ago. The Royal Nepal Academy of Science and Technology (RONAST) had prepared a detailed plan for the establishment of a science museum and a planetarium in the 1980s. Similarly His Majesty's Government of Nepal had constituted in 1992 The B. P. Koirala Memorial Planetarium Observatory and Science Museum Development Board to establish a science museum and a planetarium. Unfortunately, both these efforts proved to be unsuccessful. The

Board was reconstituted by the Ministry of Science and Technology a few years ago. The Tribhuvan University has donated some land in its premises at Kirtipur to the Board. It is learnt that the Board has prepared a master plan for the planetarium and is looking for suitable donors and other sources for funding the project.

ESTABLISHING A SCIENCE LEARNING CENTRE UNDER RONAST

The Royal Nepal Academy of Science and Technology (RONAST) is an independent apex body established to promote science and technology in the country. Since its inception in 1982, the Academy has actively been involved in building public awareness in science and technology through various fora and media. One of the functions of RONAST is to set up science centres and museums. Having its own building and land, the Academy is now considering to establish a Science Learning Centre in its premises, which will be the first of its kind in Nepal.

The overall objective of the proposed Science Learning Centre is to enhance science literacy of the Nepali populace through interactive methods. The Centre will be facilitated with models/exhibits and workshop-cum-laboratory to explain scientific principles, technological applications, and their social implications to the public in an informal and enjoyable manner. The Centre will also help preserve the country's indigenous knowledge on science and technology.

The proposed Science Learning Centre will consist of the following components:

Indoor and Outdoor Exhibition

The indoor exhibition hall will consist of interactive models/exhibits which students will see and touch, and learn by handling. The outdoor Science Park will be facilitated with all-weather type models/exhibits displayed in open space (garden) where students will learn by playing. Both types of exhibits will be prepared to explain scientific principles of various phenomenon and inquisitiveness occurring in our daily life.

Demonstration Facility

The proposed Science Center will have a workshop-cum-laboratory. The curators at the Center and/or invited science teachers will demonstrate experiments (based on school curriculum) to the students, and the students will witness the results. This will help specifically those students of the public schools with little or no laboratory facilities, understand scientific phenomenon.

School Based Activities

In order to explore and motivate young talents, the Center will organize school level competitive activities such as science exhibition, science quiz, essay writing, drawing/painting, model preparation, oratorial competition etc. These activities will form an integral part of a Science Fair and will be focused on school children.

Media Based Programs

The Science Learning Center will continue and consolidate the mass-media based science popularization programs of RONAST. These programs include publication and distribution of science features/magazines, production and broadcast of science radio programs, production of science video films, and publication of popular science series.

Mobile Exhibition

For those school children and beneficiaries, outside the Kathmandu valley, RONAST is considering to launch a special package program of mobile exhibition. Such an exhibition, accommodated in a mobile van, will consist of basic equipment and materials to demonstrate experiments, display-boards and a video show facility.

Workshops and Interaction

The Science Learning Center will organize workshops and training for the school science teachers and for the science communicators. While the teachers' workshop will provide refresher course in science teaching at school, the communicators' workshop will enhance science reporting to the public in general. RONAST will invite professionals as resource persons in such workshops. Besides the workshops, the Center will regularly organize fora like "Meet Your Scientists" where science communicators will interact with specialists on contemporary issues of science and technology.

RONAST has prepared a detailed proposal for the establishment of the Science Learning Center with technical collaboration of the National Council of Science Museums, India, with which RONAST has signed a Memorandum of Understanding for collaboration. Upon the request of RONAST, His Majesty's Government of Nepal has already submitted the proposal to the Government of India seeking its help for the setting up of the centre within the frame-work of bilateral cooperation between Nepal and India in the field of science and technology. A decision of Government of India is awaited. Establishment of the Science Learning Centre under RONAST will be a milestone in the academy's continued efforts to promote public understanding of science and technology in the country.

CONCLUSIONS

Science centres and museums have a vital role to play in enhancing public understanding of science. Although significant progress has been made in the development of science in the country, Nepal still lacks a fully-equipped science centre or a museum. In continuation of its activities to popularize science and technology in the country, RONAST is making a serious effort to establish a Science Learning Centre in its premises. Such a centre will fulfil a long-felt need and promote science literacy in the country.

ACKNOWLEDGEMENT

The author is very grateful to Dr. Dinesh Raj Bhuju and Mr. Deepak Poudyal of RONAST for their help in the preparation of this manuscript.

REFERENCES

Bajracharya, D., 2001. Science and Technology in Nepal. Royal Nepal Academy of Science and Technology, Kathmandu, Nepal.

Dwiwedi, P.K., 1976. Museums in Nepal. Kathmandu, Nepal.

Shrestha, K., 2003. Natural History Museum: Present and Future Plans. Botanica Orientalis. Central Department of Botany, Tribhuvan University, Kathmandu, Nepal, pp.153-158.

10

Museums and Science Centres in Pakistan

Dr. Muhammad Rashid Awan

Pakistan Museum of Natural History, Pakistan Science Foundation, Ministry of Science and Technology, Garden Avenue Shakar Parian, Islamabad, Pakistan

Phone 9219907, 2280889(R), Fax: 051-9221864

E-mail: awan48us@yahoo.com

ABSTRACT

Science Centers play a vital role in creating science culture in a society. They provide information as well as recreation to the students and public under one roof. Pakistan Science Foundation (PSF), an autonomous body under the Ministry of Science and Technology (Government of Pakistan) , has established a Science Centre in Faisalabad. Two Science Corners, one each at Children's Library Complex, Lahore, and Science Caravan Office, Quetta, have been established. PSF is planning to enhance this activity by establishing more Science Centers/ Corners in other parts of the country. Expertise is also offered to other organizations in establishing Science Centers. Twenty-seven Museums are located in four provinces of Pakistan: 1 in Baluchistan, 4 in NWFP, 12 in Punjab, 2 in Islamabad and 8 in Sindh. The Pakistan Museum of Natural History, Islamabad, Pakistan Science Foundation, Ministry of Science and Technology, is a unique organization in the country. PSF has established a unique program of Mobile Science Exhibition, i.e. Science Caravan. Other activities include National Science & Technology Fairs, Science Documentaries, Donation of Laboratory Equipments, Popular Science Lectures, Establishment of Science Centers/Corners and Science Essay & Poster Competitions etc. Science centers and museums are unique resources for informal education. These are the places to discover, explore and test ideas about the natural world engaging the public with science and technology. Science centers are places where people of all ages can learn at their own pace, engage their curiosity, use their senses to ask and answer questions, and explain to others what they have learned. Science centers and museums play an increasingly important role in bringing science and technology to the public.

COUNTRY PROFILE

Country Name

Islamic Republic of Pakistan

Area

887,700 sq km (550,374 sq mi)

Population

144.8 million

Capital

Islamabad.

People

56% Punjabi (also Pathans, Baluch, Mohajir, Sindhis)

Language

Urdu (also Punjabi, English, Sindhi, and regional dialects)

Religion

97% Muslim, 3% Christian and Hindu

INTRODUCTION OF PAKISTAN

The Islamic Republic of Pakistan is located in South Asia. It is a country of over 144 million population and an area of 887,700 Sq. Kms. It is blessed with abundant natural resources. A unique country, where altitude ranges from (0 meters) coastal areas to mountains (2 8,611 meters). Arabian sea is in the south, India in the east, Iran in the west, while Afghanistan lies in the northwest. China is situated towards the northeast.

Pakistan has three significant mountain ranges namely Hindukush, Karakurum and Himalaya. Hindukush mountain range extends from northeast to the southwest upto Koh-Safed. These mountain ranges contain some of the highest peaks of the world *i.e.* K2 (8611m), Nanga Parbat (8126 m), Rakaposhi (7788 m) and Tirich Mir (7690 m). Major rivers are Indus, Jhelum, Chanab, Ravi and Sutlaj. The world's largest system of interconnected agricultural canals, joins the Indus before it discharges into the Arabian Sea.

The northern and western areas of Pakistan are mountainous. Northern areas of Pakistani have some of the highest mountains in the world, including the second tallest, K-2. Northern Pakistan tends to receive more rainfall than the southern parts of the country, and has some areas of preserved moist temperate forest. In the southeast, Pakistan's border with India passes through a flat desert, called the Cholistan or Thar Desert. West-central Baluchistan has a high desert plateau, bordered by low mountain ranges. Most of the Punjab, and parts of Sindh, are fertile plains where agriculture is of great importance

Climate

Mostly hot, dry desert; temperate in the northwest; arctic in the north.

Natural Hazards

Frequent earthquakes, occasionally severe especially in north and west; flooding along the Indus after heavy rains (July and August).

Environment

Current issues: water pollution from raw sewage, industrial wastes, and agricultural runoff; limited natural fresh water resources; a majority of the population does not have access to potable water; deforestation; soil erosion; desertification. *International agreements*: party *to:* Biodiversity, Climate Change, Desertification, Endangered Species, Environmental Modification, Hazardous Wastes, Law of the Sea, Marine Dumping, Nuclear Test Ban, Ozone Layer Protection, Ship Pollution, Wetlands.

Terrain

Flat Indus plain in east; mountains in north and northwest; Baluchistan plateau in west

Elevation Extremes

lowest point: Indian Ocean 0 m

Highest Point

K2 (Mt. Godwin-Austen) 8,611 m

Natural Resources

Land, extensive natural gas reserves, limited petroleum, poor quality coal, iron ore, copper, salt, limestone etc

Land Use

Arable land: 27%; permanent crops:1%; permanent pastures: 6%; forests and woodland: 5%; other 61% (1993 est.)

Irrigated Land

171,100 Sq. Kms. (1993 est.)

Pakistan has the world's sixth largest population. This, coupled with a high growth rate, means that Pakistan is expected to overtake other nations in population in the near future, and may become the third-most populous nation by 2050 if population-control measures fail. The majority of the people of Pakistan are Sunni Muslim, with a sizeable minority of Shiite Muslims. A small minority of non-Muslims exists, mostly Christians, Hindus, and smaller groups of Buddhists and animists in the remote Northern areas.

Urdu is the national language of Pakistan and English is the official language. English is used in government and corporate business, and by the educated urban elite. Public universities use English as the medium of instruction. Urdu is the lingua franca of the people. Besides these, nearly all Pakistanis speak mutually related Indo-European languages, of which the most widely spoken is Punjabi, followed by Sindhi, Pashto and Balochi. Punjabis compose the largest ethnic group in the nation. Other important ethnic groups include: Sindhis, Pashtun, Balochis, and Muhajirs. There are also sizeable numbers of other immigrant groups such as Bengalis that are concentrated in Karachi.

Pakistan has a very rich cultural and traditional background going back to Indus Valley Civilization, 2800 BC–1800 BC. The region that is now Pakistan has in the past been invaded and occupied by many different peoples, including White Huns, Persians, Arabs, Turks, Mongols and various Eurasian groups. Thus modern Pakistani culture has its origins in the mixture of many cultures. There are differences in culture among the different ethnic groups in matters such as dress, food, and religion, especially where pre-Islamic customs differ from Islamic practices.

SUBDIVISIONS OF PAKISTAN, DISTRICTS OF PAKISTAN

Pakistan has 4 provinces namely Baluchistan, North-West Frontier Province (NWFP), Punjab and Sindh. The provinces are further subdivided into a total of 105 districts.

MUSEUMS AND GALLERIES IN PAKISTAN

At the time of independence Pakistan inherited seven museums, out of which Museums at Lahore and Peshawar were run by the Provincial Governments while site museums at Harappa, Moenjodaro, Taxila and the Armoury Museum in Lahore Fort were looked after by the Central Government through its Department of Archaeology. Immediately after independence plans were taken in hand to establish the National Museum of Pakistan. It was on 17th April, 1950 that the National Museum of Pakistan was inaugurated in Freer Hall building, replacing the defunct Victoria Museum. In order to advise the National Museum of Pakistan to enrich its collection through new acquisitions, an Advisory Council for the purchase of antiquities and works of Arts was also formed by the government in 1950. The development of museums in the country has been spectacular and the last decades have seen quite a few museums being established and many others reorganized on modern lines. Recently a museum exclusively devoted to Allama Iqbal at Javed Manzil, Lahore, and another House Museum in the Quaid-I-Azam House (Flag Staff House), Karachi, have been established. The details of the existing museums under the charge of the Department of Archaeology & Museums as well as some of the private museums are given below:

Provinces:	*Sl.No.*	*Name of Museum*	*Establishment*
Baluchistan:	1.	Quetta Museum, Quetta	1972
NWFP:	2.	Peshawar Museum	1947
	3.	University Museum, Peshawar	1995
	4.	Chakdara Museum	1960
	5.	Archaeological Museum, Saidu Sharif, Swat.	1963
Punjab:	6.	Archaeological Museum Taxila, Distt. Rawalpindi	1990
	7.	Armoury Museum, old Fort, Lahore	1928
	8.	Mughal and Sikh Galleries in Lahore, Fort	1928
	9.	Archaeological Museum, Harappa, Distt: Sahiwal	1965
	10.	Allama Iqbal Museum, Javed Manzil, Lahore	1967
	11.	Museum-cum-Library at Iqbal Manzil, Sialkot	1977
	12.	Bahawalpur Museum, Bahawalpur	1977
	13.	Faqirkhana Museum, Lahore, (Private)	
	14.	Chughatai Museum, Lahore, (Private)	
	15.	Shakir Ali Museum, Lahore, (Private)	
	16.	Arts & Crafts Museum Lahore, (Punjab govt.)	
	17.	Science Museum, Lahore	
	18.	Lokvirsa Museum, Islamabad	
	19.	Pakistan Museum of Natural History, Islamabad	
Sindh:	20.	National Museum of Pakistan, Karachi	1990
	21.	Museum at Quaid-I-Azam Birth Place, Karachi (Flag Staff House Museum)	1950
	22.	Archaeological Museum, Moenjodaro Distt. Larkana	1953
	23.	Archaeological Museum, Banbhore, Distt. Thatta.	1967
	24.	Archaeological Museum, Omerkot, Distt. Tharparkar	1967
	25.	Archaeological Museum, Hyderabad Fort, Hyderabad	1968
	26	Moenjodaro Museum, Moenjodaro	
	27	University Museum, Hyderabad	
Galleries			
Karachi		Arts Council Gallery, Ali Imam Gallery, Koonj Gallery	
Lahore		Al-Hamra Art Gallery, Lahore Art Gallery, Croweater Gallery	
Islamabad		National Art Gallery, Nomad Art Gallery, Rohtas Art Gallery (Lok Virsa) Folk Heritage Museum, Islamabad.	

SCIENCE POPULARIZATION ACTIVITIES

Popularization of science is one of the statutory functions of Pakistan Science Foundation. It has also been emphasized in the National Science and Technology Policy. The Foundation is the premier organisation that has undertaken a number of science popularization activities since its inception. Major aim is to increase awareness among the masses about the role played by science in the development of a nation. In order to extend the concepts of scientific phenomena to the majority of our population living in the rural areas of the country, the Foundation has established a unique program of **Mobile Science Exhibition**, i.e. **Science Caravan**. Other activities include National Science & Technology Fairs, Science Film Shows, Donation of Laboratory Equipments, Popular Science Lectures, Establishment of Science Centres/Corners and Science Essay & Poster Competitions etc. Brief description of some important programs is as under :

Science Caravans

Science Caravan is a mobile science exhibition that has been especially designed for rural areas. The purpose is to motivate the younger generation of Pakistan towards the study of science. Through the mobile science exhibition, the people living in rural and backward areas of the country are exposed to some of the most fascinating scientific and technological developments of the modern world. All narrations are in the national language, and are accompanied by simple illustrations. At present, five Science Caravan Units are operating in Baluchistan, Sindh, NWFP, Punjab and Federal Areas. Establishment and preparation of four new Science Caravans is in its final stages.

National Science & Technology Fairs

The main objective of the fairs is to bring the general public face to face with the technological development of the country. Four National S & T Fairs have been organized so far. Students of various levels take keen interest in such fairs due to their various interactive events. Major activities of the fairs are as under:

- Stalls of all S&T Organizations of the country
- Symposium
- Walk & Race for Science
- Book & Computer Exhibitions
- Aero modeling Competition
- S&T Art Competition
- Science Models Exhibition/ Competition

Donation of Laboratory Equipment

Under this program, the Foundation chooses the schools for donation of laboratory equipment in consultation with the Directorates of Education. Government High Schools having laboratories, science students and teachers but lacking equipment are considered for donation. Complete sets of Laboratory Equipment required as per syllabus is being provided to four-five schools each year under this program. Computers along with printers are also provided to the selected schools.

Intra and Inter Board Science Essay & Poster Competitions

This is a regular & very interesting program which is being carried out in collaboration with all (22) Boards of Intermediate & Secondary Education of the country. Fourteen competitions have been organized up till now. Topics relating to various scientific themes and issues are communicated for essay and poster competitions to the Boards for organizing Intra Board Competitions. Final competition i.e. Inter Board Contest is held in the Foundation. Cash prizes are awarded to all the winner students of these competitions.

Establishment of Science Corners/Centres

Science Centres play a vital role in creating science culture in a society. They provide information as well as recreation to the students and public under one roof. The Foundation has established a Science Centre in Faisalabad; two Science Corners, one each at Children's Library Complex, Lahore, and Science Caravan Office, Quetta, have also been established. The Foundation is planning to enhance this activity by establishing more Science Centres/Corners in other parts of the country. Expertise is also offered to the organizations in establishing Science Centres.

Other Science Popularization Programs

- Preparation of scientific brochures, booklets and posters
- Science films and planetarium shows
- Popular science lectures
- Donation of science magazines to schools
- Science quiz competitions
- Summer/winter schools
- Donation of science related books
- Sponsoring science popularization activities and publication of books
- Organizing World Science Day for Peace & Development and Foundation's Day

MESSAGE BY THE CHAIRMAN

Pakistan Science Foundation has had the privilege of serving the nation for about thirty years. During these years, PSF has contributed tremendous efforts for the promotion and funding of scientific activities.

The role of two subsidiary organizations of PSF, *i.e.* Pakistan Museum of Natural History (PMNH) and Pakistan Scientific and Technological Information Centre (PASTIC) are worth mentioning.

Pakistan Museum of Natural History is the repository of natural history of the country. Through this vast national resource, a Ph.D programme is being launched through Pakistan Institute of Natural Sciences (PINS) within PMNH.

Pakistan Scientific and Technological Information Centre (PASTIC) is responsible for collecting and disseminating scientific and technical information. It is in the process of establishing the National Science Library and Database. It strives to be the gateway of information for our scientists and industry.

We must strive for commercialization and application of our technological capabilities because no country can advance without the appropriate use of its indigenous technologies. PSF has taken up the task to make our country self-sufficient in this most vital resource of our times and hopefully this objective will be accomplished.**(Dr. Farid A. Malik)**

PAKISTAN SCIENTIFIC & TECHNOLOGICAL INFORMATION CENTRE (PASTIC)

URL: www.pastic.gov.pk

E.mail: pastic@isb.pol.com.pk

Objectives

To serve as a gateway for scientific & technological information for R&D, enhancing access to and delivery of information by catering to the information needs of researchers, entrepreneurs, industry, educationists, policy makers and planners through anticipatory and responsive information services.

Mission

Pakistan Scientific & Technological Information Centre (PASTIC) is the premier organization in the field of scientific and technological information handling and dissemination, serving thousands of researchers in the country. It is one of the first public sector organizations to get ISO–9001 certification. PASTIC is a sister organization/unit of the Pakistan Science Foundation (PSF), an autonomous organization under the Ministry of Science & Technology. PASTIC National Centre is housed in its own building at Quaid-e-Azam University Campus, Islamabad, having sub-centres at Karachi, Lahore, Peshawar, Quetta, Faisalabad and Muzaffarabad.

PASTIC is dedicated to fulfilling the ever-growing information needs of the scientific community in the fields of basic & applied sciences.

PASTIC has been designated as the National Focal Point (NFP) of Inter-governmental Council of UNESCO for the Information for all Programme (IFAP), WHO/CEHANET, SAARC Documentation Centre (SDC), UNESCO/ASTINFO, UNEP/INFOTERRA and ECO.

Objectives

- To procure, process and disseminate scientific and technological information to the researchers.
- To interact with regional and international information agencies/networks.
- To develop inter-library cooperation and resource sharing at the national level.

- Human Resource Development in contemporary techniques and methods of information handling.
- To develop and strengthen the National Science Reference Library.
- To provide bibliographic services.
- To print and publish S&T literature and provide reprographic services to S&T organizations.
- To develop Management Information System.

Achievements

- ISO Certification
- Human Resource Development
- Digital library
- Digitalization of Document Delivery Service
- Establishment of Sub-centres at Faisalabad and Muzaffarabad

Activities & Information Services

Bibliographic Information Service

PASTIC undertakes literature surveys on specific topics on request from researchers from the following international bibliographic databases containing abstracts on CD-Rom on Life Sciences, Health & Medicine, Biotechnology, Engineering & Applied Sciences, Environment, Physics, etc. E-mail: sbo@pastic.gov.pk

Document Supply Service

PASTIC acts as clearing house for scientific literature for the scientists. Copies of research articles/reports, conference papers, patents and other scientific documents published abroad or in the country are provided to the scientists/research workers on demand under this service. For this purpose, a search is made for availability of the required literature in Pakistan and if the same is not available locally it is procured from aboard. This service is provided at a highly subsidized rate.

Technological Information Promotion System (TIPS)

TIPS is a specialized network that provides support and information services in the areas of technology such as, product development, finance, economics, technical and quality regulation, manufacturing of environmentally safe products, to business enterprises.

Patent Services

Most researchers in the country lack awareness about the patent system and its benefit to the individuals as well as the society and economy. The situation needs to be improved through various promotional activities and efforts. The foremost in this endeavor is to create awareness amongst the scientific community about the benefits of inventions and patents so that research is oriented towards obtaining patents. Moreover, they should be informed as to how to get their inventions/research patented, since the level of awareness regarding filing of patents is also very low. The newly launched Patent Information Services are focused on providing advice on technical and financial matters and guiding the researchers/inventers in filing their applications for patents and obtaining protection for their inventions and research.

Entrepreneurship Development Services

PASTIC has established an Entrepreneurship Development Centre at the National Centre and one each at all the PASTIC sub-centres. Through these services the Centres will provide guidance, training and the information tools that entrepreneurs need to grow their businesses. They will learn to develop entrepreneurial skills.

Publications

Pakistan Science Abstracts

Union Catalogue

Directory of Scientific & Technical Periodicals of Pakistan

Bairun-e-Mulk Tijarat

White Meat Journal

Facilities

Computer & IT Lab.

National Science Reference Library

Reprographic Unit

Development Activities

On-going Projects prior to 2003-2004.

Establishment of National Science and Technology Database/Information Network at PASTIC

Objectives

- To strengthen the S&T information service.
- To strengthen resource sharing
- To strengthen national science reference library service
- To promote documentation service including abstracting and indexing services.
- To promote scientific and technological information transfer services
- To build up better sources for preparation of bibliographies and information dissemination services
- Acquisition of Full Text CD-ROM/Internet Databases for strengthening the information sources of the library
- Library automation.

Science Awareness Through T.V.

Objectives

- To motivate the youth of the country to learn about science and technology.
- To demonstrate science concepts using science films, video clips, computer animation and panel discussions.
- To reinforce concepts learned in the TV shows with telephonic and e-mail interaction and hands-on follow-up activity group use.
- To bring the excitement of science learning to lay audiences everywhere.
- To enhance the capability of scientists and researchers for presenting their point of view through the mass media.

New Projects Submitted

Research grants for filing of patents and publishing of articles.

Objectives

- To enhance research activities by patenting the indigenous research nationally and internationally so that the benefits of that research will reach the grass root level as well as our scientists.
- To strengthen the S&T activities by giving financial support to our scientists to publish their work in different reputed journals around the world. This programme will introduce our research and researchers to the scientists of other countries.

Strengthening of Technological Information Services at PASTIC

Objectives

- Generation of innovative activity, encouraging inventions in the country.
- Facilitating the flow of information from S&T sector into the industrial sector.
- Awareness of IPR especially patent system in the country through Workshops and Seminars organized with collaboration of the Pakistan Patent Office and WIPO.
- Development of database of Pakistani patents and dissemination of patented technological information in Pakistan.
- Collaboration with Patent Office for filing of patents.
- To develop collaboration with SMEDA, Pakistan Industrial Technical Assistance Centre (PITAC), Pakistan Intellectual Property Rights Association (PIPRA), WIPO and STEDEC for awareness and effective development of Patent Information System in Pakistan.

Enhancement of Documentation Services of Pakistan Scientific & Technological Information Centre (PASTIC) and the Establishment of Full Text National Science & Technology Database (NSTD)

Objectives

- To strengthen the S&T information service by international liaison and resource sharing amongst national organizations. A prompt documentation delivery system is needed and that is only possible by digitization and making virtual National Science Reference Library.
- To enhance the documentation services by establishing a full text S & T National Data Base concerning the indigenous research at our universities/ research organizations and research journals published in Pakistan and connectivity of digital library of PASTIC National Centre with its six sub centers for exchange of databases/S&T information.

ADRESSES OF PASTIC NATIONAL CENTRES AND SUB-CENTRES

PASTIC National Centre

Quaid-e-Azam University Campus,

P.O. Box 1217, Islamabad (Pakistan)

Tel: 92(051) 9201340,

Fax: 92(051) 9207211

E.mail: pastic@isb.pol.com.pk

Karachi Sub-Centre

A-43, Block No.3, Gulshan-e-Iqbal, Karachi.

Tel: 92(021)4987851

Fax: 92(021) 4987851

E.mail: pastic@cyber.net.pk

Lahore Sub-Centre

PCSIR Laboratories Complex,

Ferozepur Road, Lahore.

Tel: 92(042) 5756827

Fax: 92(042) 5877433

E.mail: pasticlh@pol.com.pk

pasticlh@hotmail.com

Quetta Sub-Centre

Al-Murtaza House, Near Sada Bahar Bus Terminal, Sariab Road, Quetta.

Tel: 92(081) 9211710

E-mail: pastic_ay@yahoo.com

Peshawar Sub-Centre

PCSIR Laboratories, Palosi Road, Peshawar.

Tel: 92(091)5704042

E-mail: pasticpsh@hotmail.com

Faisalabad Sub-Centre

Faisalabad Science Centre, Jail Road, Faisalabad.

Tel: 92(041) 614976

E-mail: pasticf@yahoo.com

Muzaffarabad Sub-Centre

Kashmir Information Resource Centre,

University of Azad Jammu & Kashmir, New Campus, Muzaffarabad.

Tel: 058810-44086

Fax: 058810-44102

11

Proposed National Norms and Standards for A Network of Science Centres in South Africa

Mr. K.M. Lebs Mphahlele

Manager Science and Youth, Department of Science and Technology, South Africa, Post Bag X 894, Room 6108, 188 Schoeman Street, Pretoria, 0001

Phone: +27 12 3174391/0 (O), 8622925, E-mail: lebs.mphahlele@dst.gov.za

ABSTRACT

The White Paper on Science and Technology of 1996 advocates a "campaign to promote awareness and understanding of Science and Technology and of its importance". This campaign commenced with the Year of Science and Technology in 1998 and focused on two goals: viz. to promote (a) science and technology literacy; and (b) the power of science and technology. The momentum of the Year of Science and Technology was maintained through the incremental implementation of the National Science Week with effect from the year 2000. Activities of the National Science Week have largely been effectively implemented at and by the science centres. In order to explore the potential of science centres to realize the goals outlined by the Science and Technology White Paper, the Department of Science and Technology commissioned two studies: (a) the survey on science and technology centres by the HSRC in 1999 and (b) the feasibility study on a network of arts and science centres by Ulwazi in 2004. Though appreciative of the immense potential of science centres, these studies found science centres in South Africa to be unevenly distributed and varied in terms of definition, ownership, size and capacity. Most were found to have been in operation for less than 5 years and, collectively, had attendance of about 518 000 people in 2003. Few were evaluated, prepared annual reports with comprehensive budgets and actively marketed themselves. The Department has since 2001 been providing financial support for operational, programmatic and infrastructure costs of a number of science centres. This form of support is intended to maintain the science centres while finalising the norms and standards for a network of science centres in South Africa. This paper discusses the thrusts of the proposed norms and standards with the view to sharing with and seeking ideas from the international community, especially the developing world, as they grapple with ways of using this infrastructure to support science, technology, engineering and mathematics education.

INTRODUCTION AND BACKGROUND

The Department of Science and Technology identified a network of science centres as an important infrastructure required to achieve the science and technology goals identified by the White Paper on Science and Technology (1996) and the Research and Development Strategy (2002). In order to inform the development of an effective infrastructure that will be used to implement strategies of the Department to achieve the above-mentioned goals, the then Department of Arts, Culture, Science and Technology commissioned a study in 1999 to 'take stock of what was happening in science centres in South Africa'. The study found the following about science centres in South Africa:

- They were unevenly distributed.
- Their main purpose was to support schools and develop human resources.
- In general they focused on school children and teachers.
- They largely existed in the following two forms: those with interactive exhibits and those with little or no exhibits but supported teaching and learning.
- Most lacked capacity to fulfil their roles.
- They were under-staffed and heavily dependent on part-time or volunteer staff.
- In general they were closed on weekends.
- Most were not run in line with basic business principles.
- None had had any form of external evaluation.

The study recommended a broad Government policy to support science centres to be co-ordinated centrally under a representative umbrella body linked to education at both national and provincial levels. The proposed body should be Government-funded and officially mandated to carry out its duties and raise additional funding from the private sector.

The study also recommended an impact assessment study to determine the impact of science centres and their roles in society. Between 1999 and 2004, in-depth discussions were held on the role of science centres in South Africa. During this period the Department was involved in providing ad-hoc support to a small number of science centres.

In order to formulate a well-informed policy framework for a network of science centres in South Africa, the Department of Science and Technology commissioned in February 2004 a feasibility study. The study had the following four phases:

(1) The audit phase
(2) The modelling phase
(3) The consultative and analytical phase and
(4) The recommendations phase

Data collected during the feasibility study included literature review, interviews, questionnaires, personal visits to science centres in South Africa and abroad, ten workshops in nine provinces and a national conference where draft norms and standards were tabled and approved.

In general the findings of the feasibility study confirm those of the descriptive study of 1999. Furthermore the study found the following:

- In the past five years science centres appear to have more than doubled in number.
- Most of the science centres were found to be less than five years old.
- Only 40% of the operating centres kept attendance records.
- Most centres mainly focus on general science.
- They have relatively few exhibits (66% had fewer than 50 exhibits while only four had more than 200 exhibits).
- They generally have under-utilised facilities and capacities.
- They have variable staff complement (ranging from 1 to 24 full-time staff and about 190 volunteers).
- They are variably owned (mostly by universities and parastatals and to a lesser extent, the private sector) and located (in museums, educational institutions, shopping centres, parks, research facilities, entertainment complex and community complexes).
- Competences are uneven (with fundraising and marketing being the poorest and general management and education being the highest).
- Reporting is generally poor.
- Assessment and evaluation is largely non-existent.
- Sustainability is a major problem.

The above findings were presented in ten workshops conducted in all the nine provinces during the June-July 2004 period. Represented at these workshops were more than 70 key stakeholders by more than 200 participants. Although participation was not representative of the demographics of the provinces, valuable proposals were made and collated into the proposed norms and standards document that the national conference adopted on 22-23 August 2004.

All provincial workshops conducted during the Phase 3 of the feasibility study overwhelmingly supported the establishment of a network of science centres in South Africa. The network was seen as critical infrastructure required to address the shortcomings of both science, technology, engineering and mathematics (STEM) education system.

THE VISION

South Africa inherited a system that produces pitifully low number of graduates in science, technology, engineering and mathematics; especially among those from the historically disadvantaged backgrounds.

The post 1994 Government prioritised science and technology as key to socio-economic development and established a dedicated Ministry and Department of Arts, Culture, Science and Technology (DACST). The major achievements of DACST include the declaration of the White Paper on Science and Technology in 1996, the implementation of the Year of Science and Technology in 1998 and implementation of the National S&T Week from 2000.

In 2002, Cabinet approved the National Research and Development Strategy which identified 'frozen demographics' as a major challenge facing the national science system. It is against this background that Cabinet approved separation of DACST into two distinct departments: viz. the Department of Science and Technology, and the Department of Arts and Culture. In 2004, the President appointed two new Ministries for Science and Technology, and Arts and Culture respectively.

The vision and mission of Science and Technology are to 'create a prosperous society that derives enduring and equitable benefits from science and technology' and 'develop, coordinate and manage a national system of innovation that will bring about maximum human resource capital, sustainable economic growth and improved quality of life for all' respectively.

A network of science centres is one component of a broader Departmental strategy to achieve the above vision and mission statements. The support for the use of this infrastructure is also part of a global strategy to use more than 1200 science centres worldwide which employ staff of more than 100 000, attract about 250 million visitors and inject into the economy about R10 billion, to boost the number of youth (especially those from disadvantaged backgrounds) who pursue careers in science, technology, engineering and mathematics.

GOALS

The Research and Development (R&D) Strategy and other Government Strategies such as the Human Resource Development (HRD) (DoE & DoL, 2000) Strategy and the National Strategy for Mathematics, Science and Technology Education (NSMSTE) (DoE, 2001) identified the following as key factors impacting on the development of adequate human resource capacity in the fields of mathematics, science, engineering and technology: (a) vicious under-supply of matriculates with quality passes in mathematics and science, (b) lack of financial support for those who wish to enrol for further studies in science, engineering and technology (SET) careers, and (c) lack of and inadequate facilities and resources to enhance effective teaching and learning.

To contribute towards addressing some of the above, the national conference on a network of science centres supported the following as key goals to be achieved by the envisaged network of science centres:

- To promote science literacy among the youth and the population in general.
- To enhance learner participation and performance.
- To provide career education among the youth.

Although all science centres should, as a minimum requirement, achieve the above three broad goals, specific local objectives such as tourist attraction, improvement of inner city decay, resuscitation of depressed areas, rural development and urban renewal could be accommodated in the network.

Definitions

The studies conducted by the Department of Science and Technology found that the concept of a science centre is variedly interpreted. The 1999 study describes "science centres as collections of interactive science exhibits in galleries or exhibitions at more traditional museums, each of which is designed to represent an idea or concept". The feasibility study of 2004 found that internationally a science centre is defined as a "permanently established educational facility that offers an informal educational experience in science, technology, engineering and mathematics (STEM) by providing interactive exhibits and displays and interactive programmes".

The above definitions excluded the so-called science resource centres that mainly support teaching and learning through facilities such as libraries, computer rooms and career education centres. The above definitions were, therefore, found to be narrow and limiting as they could not address the main challenges facing mathematics, science and technology education in South Africa; viz. increasing enrolments and quality passes in mathematics and science.

The provincial workshops and the national conference adopted the following inclusive and broader concept of a science centre: "a science centre is a permanently established education facility that provides an interactive educational experience through the use of interactive science, technology, engineering and mathematics exhibits, displays and programmes".

In broadening the definition the notion of 'interactivity' was unpacked and understood to cover exhibits and displays that can be handled and require a response, which ideally should be manual, mental, emotional and social. Given the above definition of interactivity, interactive exhibits could include the environment, botanical gardens etc. Interactive exhibits and displays would be used to excite and entertain as well as link educational experience to learning outcomes. Care must thus be taken in the design and development of exhibits to cater to the needs of all those who visit or use the science centres, especially those from disadvantaged backgrounds and the disabled.

A network of science centres, therefore, would refer to a group of science centres that are interconnected, aligned to and supported by the Department of Science and Technology. Membership to such a network would be voluntary but satisfy the set membership criteria. Voluntary networks that do not fit the above definition would thus not be part of the DST network.

In the context of a network of science centres, "science" shall include natural sciences, biological sciences, life sciences, technology, engineering and mathematics.

The envisaged network of science centres needs to accommodate and include a variety of science centres in its ambit if its broader goals are to be achieved. This includes science centres with specialist and generic designs, thematic and discipline-specific designs, mobile and stationary science centres, and new and established science centres. This classification enables a continuum of different variations and forms of science centres to exist in the network.

Principles

In order to 'level the playing field' described by the two studies (DST, 1999 and 2004), a network of science centres in South Africa shall be underpinned by the following principles:

- Access (centres should be strategically positioned for easy access by its target audience)
- Equity (the model adopted should be such that all benefit from the interactive educational experience offered by science centres)
- Redress (the cycle of mediocrity should be broken by ensuring that the historically disadvantaged are increasingly represented in staffing and SET human resources development programmes)
- Key performance indicators-driven delivery (through formal evaluation, centres' contribution to DST's goals will be measured)
- Synergy (synergy of efforts will be ensured to minimise duplication and negative competition)
- Sustainability (centres will strive for less dependency and building into their plans income-generating programmes)
- Ownership (centres will ensure community participation, involvement and ownership)
- Inclusivity (inclusive education will be promoted by not discriminating against any citizen on the bases of physical disability)
- Outcomes-based Education (the centres' programmes should be outcomes-based by design)
- Indigenous Knowledge Systems (indigenous knowledge systems will underpin all programmes offered by the network of science centres)
- A science centre must use a defined formula to serve a critical mass of people and schools (see norms and standards for size, location and attendance below).
- There must be a demonstrable political buy-in by provincial/local leadership for the establishment of a science centre in an area.

- A science centre must demonstrate its ability to market itself and manage its finances.
- A science centre must also demonstrate its ability to expand its services should the need arise.

Geographic Size and Space

The geographic space of science centres ranges between 50 and 4000 m^2 although some intend expanding their size in the near future. Only a few science centres indicated that their space was fully utilised although many mentioned the inadequacy of their facilities.

Most science centres offer less than 50 exhibits and very few have more than 200 exhibits and displays. Capacity to develop and renew relevant exhibits is generally low. In order to achieve the above-mentioned goals, the envisaged network of science centres shall have the following minimum size and space standards:

The network of science centres shall aim by 2020 to be able to provide about 3,75 million experiences a year to South Africa's 15 million school children.

- Normatively, a full service centre of 1 800 m^2 size should be able to cater for 50 000 school visitors in groups.
- A community of 200 000 students deserves to benefit from being served by a full service centre, subject to this centre being within easy reach (say) 100 kilometres. Depending on their location, such centres may also be able to attract up to an equal number of other visitors (e.g. children out of school, family groups and tourists).
- A limited service centre of 600 m^2 size should be able to serve 25 000 children a year in school groups.
- A mobile centre should be able to reach 50 000 students a year, focusing on rural and remote communities.
- The number and types of interactive exhibits and displays shall correspond with quality interactive educational experience and achievement of learning outcomes.
- Science centres shall make room for extensions to cater for future expansion and outdoor environment.
- A balance will be struck between quantity and quality of all the components that make up a science centre.
- Science centres shall ensure that access requirements for disabled visitors are addressed.
- Capacity of science centres will be built to achieve the set goals.

The above targets require by 2020 a three-fold increase in the number of science centres in order to achieve the above targets. It is believed that if this vision is vigorously pursued, South Africa would be able to reach its intended goal of a scientifically literate society with more of our youth pursuing careers in STEM.

Location and Ownership

Science centres in South Africa are unevenly distributed, variably located and owned. Many are owned/managed by educational institutions and parastatals. They are variably located in museums, shopping centres, office complex, theme parks, research facility, entertainment complexes, sporting complexes, community complexes and provincial parks. Only one science centre that is funded by a provincial Department of Education is freestanding.

To ensure equitable distribution of science centres in the network, the following shall serve as criteria for locating science centres:

- Location and distribution of science centres shall meet the needs of visitors.
- Each province shall have science centres in energy nodes to serve a critical mass of people, especially the youth and learners.
- Disadvantaged areas with critical mass will be prioritised.
- Local ownership including abilities and capabilities to secure and safeguard the facility shall be prioritised.

Attendance

Science centres operate between five and seven days per week while the schedule of mobile science centres and resources depends on school terms and other events. Only four centres keep a detailed attendance record. However all but one centre was able to provide total attendance for 2003 and attendance ranged from 400 to 142 000. The 20 centres including five mobile resources reached about 518 000 people in 2003–of which about 335 000 (65%) were children.

Improvements in recording attendance would help individual centres/mobiles and those seeking an overview of the sector. To achieve this objective, the network of science centres shall adhere to the following attendance requirements:

- All centres will keep attendance records of learners and experiences including those visiting or using mobile components of science centres.
- Incentives will be used to encourage attendance and promotion of higher interactive educational experience.
- A minimum standard for attendance shall be set to match the size of the centre.
- Learners will each visit a science centre at least once in their schooling career.
- All mathematics, science and technology educators will visit a science centre at least once a year in each schooling band.
- The board of governors in consultation with local communities shall determine entrance fees.

Staffing

The number of full-time personnel employed by science centres ranges from one person to 24 people per centre. The 20 centres employ a total of 159 people and 190 were mainly volunteers employed on a part-time basis.

Most centres regard their general management and education skills as being good while their administration is described as adequate and their fundraising and marketing skills as adequate to poor. Only ten centres (50%) indicated that their staff were fully utilised while most said they needed more staff.

To improve on these shortcomings, the envisaged network of science centres will use the following as criteria for staff employment and development:

- Redress and equity principles will drive employment of staff.
- Every science centre shall have a permanently tasked manager.
- The number of permanently employed staff shall depend on the budget, size and load. One educator for 10 000 visitors should serve as a guide.
- Employment of temporary staff shall depend on the ability of the science centres to attract funding.
- The options of secondment, internships and learnerships will be explored.

Financial Management and Funding

Less than half of the science centres were found to be preparing annual reports and detailed breakdown of operating income and expenditure. Income in part reflects entrance fees which range from an upper level of R 24/R20 for adults/children to a level of R5 for adults/children. Only two centres did not charge entrance fees. Overall the private sector appears to provide substantially more than the public sector toward operating income.

In general operating expenditure for salaries, maintenance and administration for science centres that form part of a museum, parastatal or university are covered elsewhere in the institution.

Due to the scarcity of recorded data, analysis of visitor attendance was difficult to obtain. It may be concluded that accurate financial information may not be important to many owners/operators of science centres. Such centres are likely to be considered only as a part of or add-on to their main activities.

Many science centres would indicate that they have never been funded adequately, that they need to scramble to secure additional support and that they are forced to curtail programming and staffing to make ends meet. However in spite of these difficulties their continued existence, often over many years, is evidence that most are largely sustainable. Nevertheless to fully assess the general viability and sustainability of science centres, it would be necessary to be assured of improvements in financial reporting.

In general financial accountability is best achieved through the services of an auditor. The majority of science centres (70%) prepare a budget and a strategic/business plan (60%). Few however conduct market surveys or produce marketing plans. Many would appreciate Government funding for operating costs and the setting up of new science centres in areas where they do not exist.

To put the network on a sound financial footing the Department will:

- Use a pre-determined formula to cover operational costs depending on the developmental stage of the centre.
- Use incentive costs to incubate new science centres and, in collaboration with private sector, provide initial 100% support to new centres and reduce the support over a period of three years.
- Use Grants-in-aid fund to support some of the projects to be offered by the network.
- All science centres shall operate on sound financial management principles in accordance with the Public Finance Management Act and the Municipal Finance Management Act.
- Encourage science centres to strive for financial self-sustainability through cross-subsidisation.
- All science centres will audit their books, develop strategic and business plans and annually report on their financial performance.
- All science centres must develop financial management measures (e.g. procurement procedures, asset management, financial policy etc.).

Monitoring, Evaluation and Research

Only 50% of science centres were found to be undertaking customer surveys, while fewer than 30% undertook internal reviews. Only one sought an independent assessment. This is an area that needs improvement and the same will be effected through:

- monitoring and evaluation annually against the set key performance indicators;
- soliciting on-going feedback from visitors; and
- conducting on-going annual impact assessments.

PARTNERSHIP AND NETWORKING

Most centres have developed partnerships with other science centres nationally and a few internationally. Fourteen of them are members of SAASTEC (Southern African Association of Science and Technology Centres). Costs preclude all but one from joining an international network.

Ten centres have links and partnerships with other institutions, usually a local university, to assist with teacher training. However links with arts and cultural organisations are limited, the majority being in the KwaZulu-Natal province and few centres involve the wider community in their operations.

- The Collaboration Agreement signed by the Departments of Science and Technology and Education provides an enabling framework within which each will serve as a lead Department in areas of its competence.
- The Department will create partnerships with key players to mobilise support and resources for sustaining the work of the network of science centres.

IMPLEMENTATION

The South African constitution demarcated the country into national and provincial competences. The Department of Science and Technology has no provincial competence and is therefore required to work collaboratively with either provinces or Departments with provincial links. Following the Cabinet decision of January 2004, the Department of Science and Technology signed a collaboration agreement with the Department of Education. The collaboration plan that emerged from this collaboration agreement outlines learner participation and performance, identifying and nurturing talent and potential, placing and supporting talented STEM students and educator development as areas in which the two Departments will work collaboratively. A network of science centres has been identified by the collaboration plan as an important infrastructure to support mathematics, science and technology education.

The Department will facilitate the establishment and maintenance of the envisaged network of science centres using a phased and incremental approach. The following steps will be undertaken in the first years of implementation:

- A national task team will develop a detailed 5-year implementation plan.
- Each science centre will appoint a management committee chaired by a full-time manager.
- The manager of each centre will report to a board of governors comprising stakeholders and role-players in a local area.
- Different levels of governance structure might be adopted for different types of science centres,
- The network will be coordinated and managed through a national consultative body that would convene planning meetings at most twice a year. This body will comprise elected representatives from the board of governors.
- The national consultative body/forum will also monitor and annually evaluate the performance of the network.
- DST will use its systems to secure and disburse funds to the network.

Sustainability

Major problems facing science centres that impact negatively on sustainability include lack of financial sustainability, low staff capacity, vandalism/theft and linguistic skills.

To be sustainable the envisaged network of science centres will:

- Receive support for operational costs from Government through a set formula.
- Promote synergy and partnerships.
- Avoid duplication of resources and efforts.
- Ensure accountability.

- Develop strategies to leverage resources to sustain science centres. Such strategies could include cross-subsidisation, official formula to determine entrance fees and the use of market-driven research to decide interventions required to meet the needs of visitors.
- Ongoing capacity building programmes.

CONCLUSION

This paper outlined the research conducted in the last ten years to determine, describe and paint the science centre landscape in South Africa. The picture drawn by these studies revealed an uneven landscape with variably located, owned, funded and managed science centres.

It also outlined the process followed in developing and adopting the proposed norms and standards for a network of science centres. This process involved facilitation of ten workshops in all the nine provinces and a national conference that considered the provincial input and adopted the proposed national norms and standards of a network of science centres. The Southern African Association of Science and Technology Centres subsequently met and formally endorsed the process and supported the proposed norms and standards for the envisaged network of science centres.

The paper discussed the thrusts of the proposed norms and standards for the envisaged network of science centres in South Africa. Currently the proposed norms and standards are going through an approval process which will, according to plan, be concluded during the month of October 2004 to enable the International Science, Innovation and Technology Exhibition planned for 1-3 November 2004 to discuss the approved norms and standards.

It is hoped that the Vietnam Conference on "The Changing Role of Science Centres in the Developing World" will inform the process South Africa has embarked on about the establishment of a network of science centres to promote science literacy, enhance learner participation and performance, and provide quality career education in science, technology, engineering and mathematics.

REFERENCES

Department of Arts, Culture, Science and Technology (1996) White Paper of Science and Technology, Pretoria (SOUTH AFRICA).

Department of Arts, Culture, Science and Technology (1999) Science and Technology Centres in South Africa: A Survey Conducted by the Human Sciences Research Council, Pretoria (SOUTH AFRICA).

Department of Arts, Culture, Science and Technology (1999) Feasibility Study for the Establishment of a Network of Science and Arts Centres in South Africa, Pretoria (SOUTH AFRICA).

Department of Education (2001) National Strategy for Mathematics, Science and Technology Education, Pretoria (SOUTH AFRICA).

Departments of Education and Labour (2000) National Human and Resource Development Strategy, Pretoria (SOUTH AFRICA).

Department of Science and Technology (2002), National Research and Development Strategy, Pretoria (SOUTH AFRICA).

12

Museums and Science Centres in Sri Lanka

Arumahennadige Piyasiri
Mechanical Engineer, National Engineering Research and Development Centre
2P/17b, I.D.B., Industrial Estate, Ekala, Ja-Ela, Sri Lanka
Phone:- 094-011-2236384, 2236284 Fax: 094-011-2236434
E-Mail:- nerdc@sri.lanka.net

ABSTRACT

Sri Lanka, an island in the Indian Ocean and located south of the Indian subcontinent, is also known as the Pearl of the Indian Ocean. The country's literacy rate is 89%, which is one of the highest in Asia. Present and past all governments elected have taken several steps to strengthen the science knowledge among people. The education system of Sri Lanka has made science and mathematics as compulsory subjects for all students from grade 6 to grade 11. After success in G.C.E. (O/L) examination, a student is free to select the subjects for higher education including science subjects. Regional and national level science competitions are held annually. Annual Presidential awards for young inventors and scientists at the national level have been organized through Sri Lanka Inventors' Commission, Institution for Engineers of Sri Lanka etc. The Ministry of Science and Technology of Sri Lanka has initiated the programme for Popularization of Science in the country through Science and Technology Centres attached to all divisional secretariats. The museums are the places to discover, explore and test ideas about the natural world. In Sri Lanka too, they are situated in all parts of the country. The National Museum situated in the capital City, Colombo, was established in 1877 having collections of antiques, national treasures and artefacts from all parts of the island. Other museums display artefacts, fauna and flora of the sea and environment, gems, a model displaying the process of gem-mining, archeological finds etc. In addition the NERDC of Sri Lanka under the Ministry of Science and Technology has established the country's first and only Science and Technology Park to popularize science among people in order to improve their creativity. The NERDC has also laid the foundation for setting up the country's first Engineering Museum to display the ancient technologies developed by our forefathers.

Keywords: *NERDC (National Engineering Research & Development Centre), Engineering, Inventions, Innovations, Province, Creativity, Vidatha (Science and Technology Centres set up by Ministry of Science & Technology)*

INTRODUCTION

A teardrop-shaped island just off the southern tip of the Indian sub-continent, less than 50 kms (31 miles) away at its closest, Sri Lanka has links with the cultures of India that stretch back beyond history. But it would be a mistake to think of it as an India in miniature.

Sri Lanka lies between 5° 55' and 9° 55' north of the equator and between the eastern longitudes 79° 42' and 81° 52'. The climate of Sri Lanka is typically tropical with an average temperature of 27°C. In the higher elevations it can be quite cool with temperatures going down to 16°C at an altitude of nearly 2,000 metres.

Mainly an agricultural country, the chief crop of Sri Lanka is rice in which the country is almost self-sufficient. Tea, rubber and coconut are also important agricultural crops, with tea being a major foreign exchange earner. Other crops of importance are cocoa and spices. Sri Lanka is also a major exporter of precious and semi-precious stones. In the last three decades tourism has emerged as an important industry.

From North to South it has a maximum length of 435 kms (270 miles) and at its widest point it measures 225 kms (140 miles), giving it a land area of 65,600 sq.kms. (25,000 sq. miles). This entire stretch encompasses beautiful tropical beaches, verdant vegetation, ancient monuments and a thousand delights to please all tastes. Bright, sunny, warm days are the rule and are common even during the height of the monsoon–climatically Sri Lanka has no off season.

Sri Lanka is considered to be a third world developing country. It became independent in 1948 from British rule. After independence the country has been ruled for only 56 years by its own people. Since 1505, Sri Lanka had been ruled by the Portuguese, Dutch and English people from time to time. This report is mainly focused on the present status of museums and science centers in Sri Lanka. To discuss more about this topic it is vital to consider the pre- and post-independence periods of Sri Lanka.

In 1505 the Portuguese arrived in *Colombo* and gained a monopoly on the invaluable spice and cinnamon trade. The Dutch were more interested in trade and profits than in religion or land, although they too unsuccessfully tried to bring the *Kandyan* kingdom under their control. The British arrived in 1796. In 1815 they became the first European power to rule the entire island. The rule of the empire moved inexorably forward as roads were built, coffee, tea, cinnamon and coconut plantations sprang up and English was introduced as the national language. By a process of peaceful constitutional evolution Sri Lanka won back independence in 1948. Sri Lanka became a Republic in 1972 with the adoption of a new constitution.

In 350 years of pre-independence period, foreign nationals have established several museums such as the National Museum, the Dutch Museum etc. in the country and the museums such as the Weera Puran Appu Museum, the Bandaranaike Museum etc have been established in the post-independence period.

SCIENCE CENTRES AND MUSEUMS OF SRI LANKA

Science centres and museums are unique resources for internal learning, a process central to forging knowledge, skills and positive attitudes. They are the places to discover, explore and test ideas about the natural world. According to an unnamed scientist "Science centres are places where people of all ages can learn at their own pace, engage their curiosity, use their senses to ask and answer questions and explain to others what they have learned". Science centres are situated at a special nexus among the community, the school and the home which gives them a unique role in the educational infrastructure, one that can propel science educational reform into the 21st century.

MUSEUMS OF SRI LANKA

The National Museum–Colombo

The National Museum is situated in Sir Marcus Fernando Mawatha, Colombo-07. This first public museum was established in Sri Lanka in 1877. Best known for its collection of antiques and objects d'art displaying the cultural heritage of Sri Lanka, it includes the national treasures and artefacts from all parts of the island. A section of the first floor houses the Puppetry and Children's Museum. It also houses a library with a collection of about 500,000 books, including very valuable and rare ones and more than 4,000 ancient palm leaf manuscripts.

The museum's paintings, dating back centuries, tell a tale of a bygone era when the pace was more relaxed. There are scenes of Sri Lanka's aborigines, the Veddahs, hunting with bows and arrows.

For researchers, the museum library has over half a million publications of interest. For children, the fascinating world of puppetry comes alive with the display upstairs. There are masks used in traditional dance, including devil dancing.

Open daily from 0900 hrs to 1700 hrs except on Fridays.

Tel: 094-11 2694366, 094-11 2694767

The National Museum of Natural History–Colombo

The National Museum of Natural History is situated in the same premises as the Cultural Museum described above. It displays the Natural Heritage of Sri Lanka. The fauna i.e. the mammals and the birds are displayed in dioramas. There are sections on Applied Botany, Geology and Fossils and the Natural Environment. In the Discovery

Room one finds the mounted skeleton of an elephant and a scale model in relief showing the topography of a part of Sri Lanka.

Open daily from 0900 hrs to 1700 hrs.

Tel: 094 11 2691399

The Dutch Period Museum–Colombo

The Dutch Period Museum in the old "Dutch House" on Prince Street, Pettah (Colombo 11), which houses this museum, was built in the latter part of the 17th century and was initially the residence of Count August Carl Van Ranzow along with five other houses of the elite. Today, the sides of the street are choc-a-block with boutiques and stores of moslem traders. The restoration of this building commenced in 1977 and was completed in 1981. This museum was opened to the public in 1982. This building embodies the unique architectural features of a colonial Dutch town house. The museum while displaying the Dutch legacy, with the artefacts viz. furniture, ceramics, coins, arms etc., also portrays facets of contemporary life and culture.

Open daily except Fridays from 0900 hrs to 1700 hrs.

Tel: 094 11 2448466

The Museum–Kandy

The Museum, Kandy, is situated behind the Temple of the Tooth, in a building of historical and architectural interest. The collections are all of the Kandyan Period (17th–18th Century).

Open daily from 0900 Hrs to 1700 Hrs.

The National Gem Museum–Ratnapura

The National Gem Museum of Ratnapura is housed in the famous Ehelapola Walawwa on the Colombo Road in Ratnapura. The museum displays to a large extent the prehistory of Sri Lanka. Ratnapura being famous for gems, the process of gem-mining is displayed through a model. Some of the artefacts displayed here, reflect the unique arts and culture of the Sabaragamuwa Province.

Open daily except Fridays from 0900 hrs to 1700 hrs.

The Folk Museum–Anuradhapura

The Folk Museum is in the sacred city close to the Archaeological Museum and has a collection of artefacts illustrative of the rural life of the North Central Province.

Open daily except Thursdays & Fridays from 0900 hrs. to 1700 hrs.

Museum–Galle

The Museum, situated within the Fort of Galle in a Dutch colonial building in Church Street, is the Cultural Museum adjoining the New Oriental Hotel. The artefacts reflect the arts and culture of the Southern Province.

Open daily except Mondays from 0900 hrs to 1700 hrs.

The National Maritime Museum–Galle

The National Maritime Museum is located within the Fort of Galle in a colonial Dutch warehouse with imposing pillars. This Museum displays the fauna & flora of the sea and the environment. Artefacts consist of preserved material and scale models of whales and fishes. Generally, all the resources of the sea are covered in this museum.

It also shows in diorama form with life size models, and the traditional methods of fishing. Some artefacts of under-water archaeology are on display. An interesting experiment is the "walk-into the sea" diorama, showing the natural coral beds, sea grass beds and deep-sea fishes. Finally, one leaves the museum seeing the causes of sea pollution, coastal erosion and methods used to combat these problems.

Open daily from 0900 hrs to 1700 hrs.

Weera Puran Appu Museum–Moratuwa

Weera Puran Appu Museum is housed at Weera Puran Appu Vidyalaya, New Galle Road, Moratuwa. This Museum exhibits items in relation to the life and times of this great hero who fought the British against unfair levy of taxes and injustice and laid down his life in defence of the country.

Bandaranaike Museum–Colombo

Bandaranaike Museum is housed in the Bandaranaike Memorial International Conference Hall, Bauddhaloka Mawatha, Colombo 07. The museum contains photographs, objects and documents portraying the life and times of the late Prime Minister, S W R D Bandaranaike. Visitors can listen to recordings of some of his famous speeches.

Open daily except Mondays and full moon days from 0900 hrs to 1600 hrs.

Tel: 094 11 2691131

Archaeological Museums

The Archaeological Department has established archaeological site museums in Anuradhapura (ancient city), Isurumuniya, Minintale, Veheragala, Vavyniya, Jaffna, Sigiriya, Kandy, Dedigama, Yapahuwa, Paduwasnuwara, Kotte, Matara (Star-Fort), Deegavapiya and Kataragama. Archaeological museums are open daily from 0800 hrs. to 1700 hrs., but closed on Tuesdays and some of them will be closed on public holidays too.

SCIENCE CENTRES IN SRI LANKA

Vidatha Centres

Vidatha centres are the Science & Technology Centers which have been established by the Ministry of Science & Technology. The Ministry has initiated steps to set up the "Science and Technology Centres" all over the island. In the first stage, there will be science centers attached to all divisional secretariats. To date there are some 10 centres established in some divisional secretariats such as Badulla, Kandy, Hambantota etc.

In addition to the Vidatha Centres, the Ministry has arranged a special train equipped with some of the scientific and technological equipments in the compartments to display scientific concepts to the visitors to the exhibition train. This science train is parked at railway stations for sometime during which the interested people can visit the train for improving their science knowledge.

NERDC Science and Technology Park

The National Engineering Research and Development Centre of Sri Lanka was established in 1974. To celebrate the Silver Jubilee of the centre in 1999, the NERDC Science and Technology Park was set up in the same premises. The area covered by the park is 4.5 acres. The NERDC Science and Technology Park is the only one of its kind in Sri Lanka.

Aims and Objectives

The main objectives for setting up the Park were to:

- Enhance the creativity of the students and youth leading to inventions and innovations.
- Upgrade and popularize the knowledge of science and technology among the Sri Lankan public.
- Guide towards self employment opportunities.
- Improve productivity using energy efficient systems.
- Give value for domestic and industrial waste.
- Provide a window for NERDC activities at a glance.

Products/Services offered by NERDC S&T Park

To achieve the above aims and objectives, the NERDC Science & Technology Park offers following services to the visitors to the park:

- Making them aware of the
  - History of the NERD Centre.
  - Inventions/Technologies/Processes developed by Sri Lanka.
  - Advantages of inventions/innovations to individuals/organizations/groups as well as to the country; and
  - NERDC services offered to the public such as construction of cost effective buildings, energy audits, consultancy services, training programmes conducted and CME activities etc.
- Practical demonstrations of NERDC developed products and processes. In addition, the same products are also on sale to the interested parties through the Sales Unit.
- Teaching science concepts and technologies developed at NERDC.
- Helping the G.C.E advanced level students to gather information for their project reports.

- Motivating the people for self-employment.
- Accommodation facilities to students and teachers from far away places, trainees who register for in-house NERDC training programs/seminars etc.
- Canteen facilities, resting and first-aid facilities for the visitors to the Technology Park and occupants of the Hostel.

Currently, the NERDC Science & Technology Park exhibits and demonstrates the technological products/ processes/ projects which have been designed and developed by NERDC engineers and scientists since 1974. Majority of the exhibits available in the park are in the fields of Renewable Energy, Agricultural Engineering, Civil Engineering, Environmental Engineering, and Electrical & Electronics Engineering. In addition there are exhibits which are specially designed for G.C.E ordinary level students in grade 11.

EXHIBITS

Exhibits display in the S&T Park belongs to the following Engineering Departments:

Civil Engineering

Cost effective civil construction technologies are displayed. They are:

- Construction of cost effective foundation
- Cost effective wall constructions (slip form wall construction technique)
- Construction of door and window frames from concrete
- Pre-cast pre-stress column fabrication
- Beam fabrication
- Slab construction using pre-stress trapezoidal beams
- Cost effective roof constructions
- Cost effective buildings

Renewable Energy

Department displays the following products and processes of Solar Energy:

- Wind Energy, Bio Mass Energy etc.
- Solar dryers, stills, concentrator, water heaters etc.
- Wind mill water pump, wind mill electricity generators.
- Firewood gassifiers for thermal and electrical energy generation applications, saw dust operated bakery ovens.
- Bio-gas digesters.

Agricultural Engineering

Department displays the following products:

- Paddy storage bin
- Coconut oil expellers

- Cinnamon oil distillation
- Coconut de-husker
- Parboiled paddy mill
- Raw Paddy Mill

Electrical & Electronics

Department displays the following exhibits:

- Electronic ballast
- Fan regulator
- Induction motor
- Teaching aids (mini lab)
- Computer aided teaching
- Electricity generation through solar panels

For whom: PLACE/Customer Segment

In the beginning, the main target audiences to the Technology Park were school students and teachers. Due to its popularity, all kinds of people such as students from schools, technical colleges, agricultural schools, private educational institutes, international schools, universities etc.; parents, industrialists, unemployed and underemployed have become regular visitors to the Technology Park. In addition people from non-governmental organizations, women's organizations and staff/ field officers from banks are also frequently visiting the Park.

Price

To keep the NERDC Technology Park clean and tidy, we charge the visitors:

Rs. 10/- per student and Rs. 15/- per adult

Promotion through

- Word of mouth.
- Letters mailed to schools.
- Customers who are interested in purchasing NERDC developed products.
- Television programmes, newspaper articles, exhibitions, awareness programmes etc.
- Accommodation facilities to visitors provided in the Hostel building constructed at the Technology Park premises.
- Exhibition of newly developed products by the NERD Centre.

As the visitor arrivals to the Technology Park have been increasing annually (see Table 12.I), the services offered by the Technology Park to the visitors also have been improving. The construction of Hostel facilities for the faraway schools and

Table 12.1: Visitor Arrivals since 1999 August

Year	*Visitor Arrivals*
1999 Aug-Dec	3043
2000	6813
2001	13511
2002	20938
2003	21575
2004 September	17578

newly constructed canteen facilities have contributed to the increased inflow of visitors. The Centre has not promoted the Technology Park through electronic media yet. If it is done, the visitor arrivals can be increased two to three times.

Planned Activities for the Year 2005

- To provide accommodation for the students and teachers who visit the Technology Park from far away schools
- To provide better facilities such as those of canteen, toilets etc. by which we expect to increase the number of visitors to the Park.
- To provide Hostel facilities for the trainees, resource personnel who engage with NERDC in-house training programmes.
- To provide facilities for the resource personnel/supervisors from overseas institutions to the NERD Centre.
- To sell brochures, other booklets to the visitors.
- To exhibit new projects/processes replacing older exhibits at least once in two months.

Strategies to increase the Number of Visitors to the Science and Technology Park

- Mailing a letter to all school Principals, Technical Colleges, Samurdi Groups etc.
- Promotion of the Technology Park activities through Divisional Secretariats and local government bodies.
- Promotion through paper advertisements.
- Exhibiting new projects/processes replacing old items at least once in two months.
- Presently, science education curriculum for classes 8,9,10 and 11 consists of the concepts of teaching aids and other science & technology concepts introduced at the Technology Park. We can get the maximum benefit through this opportunity to increase the inflow of students to the Technology Park.
- Construction of the first of the proposed NERDC Engineering Museums to display past achievements of our great Sri Lankan forefathers.

PROPOSED NERDC ENGINEERING MUSEUM

In addition, National Engineering Research & Development Centre has embarked on setting up of a Centre to display past achievements of great Sri Lankans as we do not have in Sri Lanka an engineering park nor an engineering museum. The establishment of a facility of this nature could be a tremendous encouragement for the visitors to the present Science & Technology Park and will be able to seek the past achievements and current developments in a practical manner under one roof.

To mark the 30th Anniversary of NERD Centre we have laid the foundation stone for construction of the Science and Engineering Museum in which the engineering and technological products and processes designed by ancient Sri Lankan people will be installed in the immediate future. Once we complete the construction of the museum building we would be able to install ancient Technologies like Bisokotuwa, Sekkuwa etc. in order to uplift people's knowledge about Sri Lankan heritage.

RESULTS AND DISCUSSION

Past Achievements

Table 12.2: Visitors to the Park in Groups in the Year 2004

Month	*School Groups*	*Other Institutions*	*Total*
January	–	2	2
February	7	3	10
March	21	4	25
April	3	1	4
May	15	10	25
June	18	10	28
July	47	6	53
August	35	11	46
September	15	5	30
Total	**161**	**52**	**223**

Table 12.3: Visitors to the Park in Groups in the Year 2003

Month	*School Groups*	*Other Institutions*	*Total*
January	3	2	5
February	15	2	17
March	34	4	38
April	21	4	25
May	12	1	13
June	17	2	19
July	51	6	57
August	20	4	24
September	18	3	21
October	19	7	26
November	6	5	11
December	4	3	7
Total	**220**	**43**	**263**

Table 12.4: Visitors to the Park in Groups in the Year 2002

Month	*School Groups*	*Other Institutions*	*Total*
January	–	5	5
February	7	2	9
March	37	3	40
April	22	5	27
May	13	2	15
June	37	4	41
July	42	3	45
August	22	11	33
September	18	2	20
October	21	7	28
November	8	3	11
December	3	1	4
Total	**186**	**38**	**224**

Table 12.5: Visitor Inflow to the Technology Park during the Year 2003

Month	*School*	*Technical Colleges*	*Others*	*No. of Visitors*
January	02	01	–	206
February	14	01	02	1248
March	33	02	02	3364
April	23	02	01	2160
May	12	–	02	1031
June	16	01	01	1277
July	47	04	–	4667
August	39	03	02	3608
September	16	01	03	1418
October	19	01	04	1707
November	6	02	03	610
December	–	03	03	279
Total	**227**	**21**	**23**	**21575**

According to the data collected, visitors to the Technology Park have increased annually. At the same time, the number of groups and other institutions visiting the Park have also increased. (See tables) This increase is mainly through word of mouth from visitors without any attempts made through advertisements. Our services to the visitors are through lectures by our demonstrators and technical staff. Practical demonstrations are carried out and science concepts are explained. At the same time the visitors are coming to the Technology Park to gather information regarding projects for their educational reports.

CONCLUSIONS

People are thirsty for science. They are eager to learn more about science. People are learning everyday and improving their knowledge. Especially in a third world developing country like Sri Lanka, there are talented, skilled youngsters and students who can create innovative/ inventive things, if they could learn scientific and technological concepts in a practical manner during their learning process. To make people creative, the government as well as intellectuals should be able to make a national policy regarding the setting up of the museums as well as science and technology centres throughout the country to popularize science among the people. Then the people will be able to apply practical experience gained through these centres to use these skills to live better in this world.

ACKNOWLEDGEMENTS

Any project which is carried out at any level, is a result of coordination and assimilation of work done by a number of individual experts and institutions without which even to think of its completion may be compared to a dream.

Before acknowledging due help at different levels from different persons, I would like to mention the main inspiration behind the work in the name of the Chairman of NERDC, Mr. Jayantha Ranathunga for his cooperation and encouragement during the moments in which it seemed nearly impossible for me to complete the work.

My respected General Manager Mr.M.W.Leelaratne came out with his valuable instructions and suggestions from time to time for which I am very grateful to him.

I would like to acknowledge with sincere regards the help rendered to me by Deputy General Manager (Services), Mr.D.C.Jayaratne, who was never short of giving help of any kind during my work.

I am also thankful to Mrs.P.D.R.Damayanthi, Miss E.C.N.Ekanayake and Mr. K.A.J.Perera of the Technology Park for their assistance.

Last but not the least, my dear wife Bhadra and sons Shanaka, Kovida, Miyuru and Navoda and daughter Achini are the ones who count.

REFERENCES

www.a-zoftourism.com, www.namstct,org

www.srilankatourism.org/museums.htm

13

The Changing Role of Science Centers and Museums in Sudan

Ibrahim Babiker Abdalla
Advisor, Ministry of Science & Technology,
PO Box 2904, Khartoum, Republic of Sudan
Phone: +00249 1 83 472360, Mobile: +00249 9 12151748
Fax: +00249 1 83 472362, E-mail: babiker_ibrahim@hotmail.com

INTRODUCTION

Sudan is situated in the north east sector of Africa, approximately between lattitudes 4 and 22 degrees north and longitudes 22 and 38 degrees east. The total population is estimated at 31 million as projected by the 1993 census data, with an annual growth rate of 2.5%. The country is rich in natural resources both renewable and non-renewable such as water, land, petroleum and natural gas. The main sources of water in the country are the river Nile and its tributaries, rain, and groundwater. The average rainfall in Sudan ranges from a few millimeters in the extreme north to about 1800 millimeters in the extreme south.

HISTORICAL BACKGROUND

Science centers and museums are unique resources for informal learning in almost every nation on earth. Sudan is by no means any different in this respect. The learning process is central to forging knowledge, skills, and positive attitudes.

Science centers and in museums in their modern sense started as early as the beginning of the twentieth century in Sudan, with some limited effort, to improve standards of production and quality of various commodities and raw materials, together with some routine services and quality control.

SCIENCE CENTERS

In 1900 during the colonial era a veterinary quarantine unit was established within the Military Veterinary Corporation. This unit undertook the study and treatment of some endemic diseases that attacked horses used by the colonial military.

Soon after that, in 1902, the Wellcome Tropical Laboratory was established to undertake various research activities in the fields of endemic diseases and pest control. In 1913, the military veterinary corporation was upgraded to the Central Laboratory and Research Administration. The research focus of this new Administration was the production of vaccines. A number of research stations were established thereafter. These were: The fishery research station in 1953; the animal production research stations in: Gazala Gawazat; Um Benien; Atbara; Al Shukaba in 1957; and the wild life research station in 1975. All these efforts, to strengthen the veterinary and animal production research in the country, resulted in the establishment of the Animal Wealth Research Corporation (AWRC) in Khartoum in 1995.

Parallel to the development of the Animal Research Centers in Sudan, the Agricultural Research Centers were started by a small research unit within the premises of the Wellcome Laboratory in 1902. An independent Agricultural Research Unit (ARU) was established in 1904. This new research unit was attached to the Educational Services Department in 1912. By 1919, the Agricultural Research Unit was transferred to the Agricultural Services Department. A grid of agricultural research stations was developed across the country in 1918 when the Gezira Research Station (GRS) was established. This was followed by the establishment of Tozi, Abu Naama, Hudeiba, Shendi, Shambat, Al Gunied, Kassala, New Halfa, El Obeid, Kadugli, Gazala Gawazat, Al Fasher, Dongola, and Gedarif research stations. To streamline the agricultural research activities in the country, the Agricultural Research Corporation (ARC) was founded in 1968. The ARC played a significant role in terms of both manpower training and the number of research programmes that were introduced as compared to the ARU. It was clear that all these research centers were problem oriented and their role in the development of the country at large was small.

Basically, to provide medical services for the colonial army, the Kitchener school of medicine was established in 1924.The Kitchener school of medicine was later merged into the Gordon Memorial College. This was joined with other sections of the Gordon Memorial College in 1951, to form the University College of Khartoum. In 1938, the first post-secondary school of veterinary science and the first post-secondary school of agriculture were established. These later became the faculty of veterinary science and the faculty of agriculture of the University of Khartoum. These two faculties have greatly upgraded scientific research by providing qualified personnel and by developing programmes and methods for scientific research in their respective fields.

In 1956, the country witnessed the start of a new phase in the quality and quantity of scientific research. This was due to the fact that the University College of Khartoum became a fully independent university granting internationally-recognized academic degrees and furnishing scientific researchers with more and better facilities, and with more opportunities for scholarships, both government and non-government, at universities and other institutes abroad. For sure this development had contributed to the development of the country by providing specialists and professionals in the different branches of modern knowledge. The University of Khartoum was followed by a Cairo University branch in Khartoum and which, later, during the seventies, was joined with the Omdurman Islamic University, Juba University, and Gezira University. The turn out from these higher educational institutions was far less than what had been anticipated by the policy adopted during all these years. The enrolment and graduation from these institutions had never exceeded 15000 to 20000 students.

The first organizational structuring of science and technology came in 1970. A presidential decree was passed establishing the National Council for Research, as an autonomous body responsible for formulating policy, planning, designing programmes for the applications of science and technology in economic and social development, and the coordination of national efforts in this respect. The Council was directly under the President of the Republic, and having the following terms of reference:

The Council aims at encouraging, organizing and promoting scientific research in its various aspects with a view to achieving the objectives of economic and social development within the overall framework of the policy of the country.

The Council shall, with a view to achieving its objectives, formulate a comprehensive policy for scientific research and shall direct the plans, means and methods for the execution of such policy in such a manner as to ensure efficient and effective utilization of human and other resources available at the various units of scientific research.

The National Council for Research submitted a number of plans approved by the state. The most comprehensive one was endorsed by the government to cover the period 1981–1990. The necessary funds were allocated for the implementation of the different programmes, but the outcome was modest as a result of a number of constraints. Due to lack of continuity the plan was abandoned in 1985. The NCR started operating through four specialized councils. These were:

- The Agricultural Research Council
- The Economic and Social Research Council
- The Scientific and Technological Research Council
- The Medical Research Council

In addition NCR formed several other bodies. These were:

- The Sudan Atomic Energy Commission
- The National Computers Committee
- The National Committee for Man, Environment and Development
- The National Remote Sensing Center
- The National Documentation Center

In addition to the above-mentioned committees and specialized centers attached to the general secretariat, the following centers are either affiliated to the Council or are co-operating with it:

(a) Centers Affiliated to NCR

- The Institute of Tropical Diseases
- The Medical and Aromatic Herbs Unit
- The Institute of Oceanography
- The Institute of Energy research
- The General Instrument Repair Workshop
- The Cellulose Research Unit
- The National Center for Technology

(b) Institutions Co-operating with NCR

- The Universities of Khartoum, Juba, and Gezira
- Veterinary Research Administration
- The Agricultural Research Corporation
- Food Processing Research Center of the Ministry of Agriculture, Food and Natural Resources
- The Tanning Institute of the Ministry of Agriculture, Food and Natural Resources
- The Industrial Research and Consultancy Institute of the Ministry of Industry
- The Research Section of the Meteorological Department of the Ministry of Defense
- Geological and Mineral Resources Department of the Ministry of Energy and Mining
- Building and Road Research Institute of the University of Khartoum

MUSEUMS

Museums play an effective and important role for all community individuals in terms of education, culture, amusement and tourism. Historically, the word museum is of Greek origin. It refers to the place of the free arts God which include music, painting, and all humanities.

In Sudan, the first museums and archaeology department was established in 1902. What a coincidence! The Wellcome Tropical Research Laboratory was founded the same year. In 1904, a small Archaeology Museum was established at Gordon Memorial College, where it was opened for the public for educational purposes. During the period 1910–1938 a lot of digging was done in many places throughout the country. The most important of these were: The British archeologist, Marstang, in Merowy City in northern Sudan; the American archaeologist, Harvard Rayzner, in

Karrmah, Nouri, Barkal, and Bajraweia, also in northern Sudan. Most of the discoveries were displayed in the Merowy Museum in 1925. Soon after that, in 1928, "The House of the Mahdi Khalifa" was transformed into a museum for the purpose of displaying the history of the Madieyah and Turkish era. Another museum was opened in 1930 in Wadi Halfa in the extreme north of the country to display the history of the Nubien era, but was transferred to the Capital Khartoum after the construction of the High Dam at Aswan in Egypt.

After independence, in 1956, the ethno-archaeology museum was opened to take care and conserve the history of the different ethnic groups that comprise the Sudanese nation. This was one of the most important tools that unified the nation around the common goals in spite of the diversity of cultures and historic background of the different ethnic groups. The Shikan Museum was established in 1965 in El Obeid, western Sudan, in memory of the Shikan battle which was won by the mahdi army. It also displays some history for the Senar and Kordofan state traditions. In Darfur, the Palace of the Sultan Ali Dinnar was transformed into a museum to display the traditions and history of the Darfur state. The Natural History Museum was also established at the University of Khartoum and the Birds Museum was established at the Agricultural Research Corporation headquarters at Wad Medani. A War Memorial Museum was opened in the capital Khartoum, displaying the history of all the Sudanese army and the contribution of those Sudanese who participated in the First and Second World Wars.

All these efforts culminated in the establishment of the National Archaeology Museum in 1971. The National Museum was organized in the following order: Khor Abu Angah, Old Khartoum, Al Shieheinab, The First Nubein, The Third Nubien, Karmah Monarchy, and the Christian, and Islamic civilizations. The National Museum is known to be the best place to visit due to its unique location along the river Nile Mogran area where the Blue Nile joins the White Nile, thus forming the great river Nile. Also, the richness of the displayed material attracts the attention of the visitors from both inside and outside the country

NATIONAL SCIENCE AND TECHNOLOGY POLICY STRUCTURE

Science and technology policies in Sudan were not formulated in a comprehensive sense after independence in 1956, but left to institutional and academic initiatives. Nevertheless, the successive economic plans had embodied research and technology programmes. Looking carefully at the statistics of some basic indicators such as; the Human Development Indicators as shown in Table 13.1; the Human Development Index shown in Figure 13.1; the Per Capita Gross Domestic Product shown in Figure 13.2. As far as the role of science and technology in the development of Sudan was concerned, it was clear from the earlier days that the country suffered from a high degree of technological dependency both in its hardware and software. This dependency was manifested in:

- The importation of all forms of machinery and equipment and a good deal of certain intermediate commodities and other inputs.

Table 13.1: Human Development Indicators

Development Indicator	*Highest*	*Average*	*Lowest*	*Sudan*
Longevity (LEB) (years)	77.0	66.5	50.6	55.0
Literacy (%)	98.3	75.38	48.5	53.3
School Participation (%)	89.0	64.0	39.0	34.0
Income per capita ('000 US$)	21.6	3.3	0.98	1.56
Life expectancy indicator	0.87	0.6	0.43	0.50
Education enrolment indicator	0.95	0.72	0.45	0.47
Per capita GDP	0.90	0.58	0.38	0.46
Basic growth indicator	0.90	0.66	0.41	0.47

- The high incidence of dependency on foreign technical expertise and preference for turnkey projects.
- Deficiency of the indigenous specialized contracting firms in project construction and machine erection.

Various measures had been taken in the country to counteract this technological dependency. To mention some:

- The establishment of new local research and development institutions and up-grading of the existing ones.
- Restructuring of the educational system with a strong bias toward technical educational institutions
- The establishment of the Ministry of Science and Technology.

In 1989, the National Council for Research was reorganized, after the establishment of the Ministry of Higher Education and Scientific Research (MHESR). The Minister of Higher Education and Scientific Research became the Chairman of the National Council for Research. This newly formed ministry, embarked on an ambitious plan for the expansion of higher education in the country. Within only four years, twenty four new regional government universities have been established nationwide, increasing the student intake considerably (Figures 13.3 and 13.4). Another 30 private universities and colleges were established during the same period (Figure 13.5). The Ministry of Higher Education and Scientific Research's plan also included the establishment of the National Center for Research in 1991.The new mandate for NCR was to enhance scientific applied research relevant to the requirements of the economic and social development strategies.

The government's commitment to science and technology was inbuilt into the National Comprehensive Strategy Plan (NCSP) 1992–2002 which was aimed at taking the necessary measures for responding to the need of the development projects.

The NCSP endeavours to enable the country acquire modern means of technology and their application in production and service sectors as well as educational and guidance curricula. The outcome of all of these efforts was far behind the great hopes

Figure 13.1: Human Development Index (Composite)

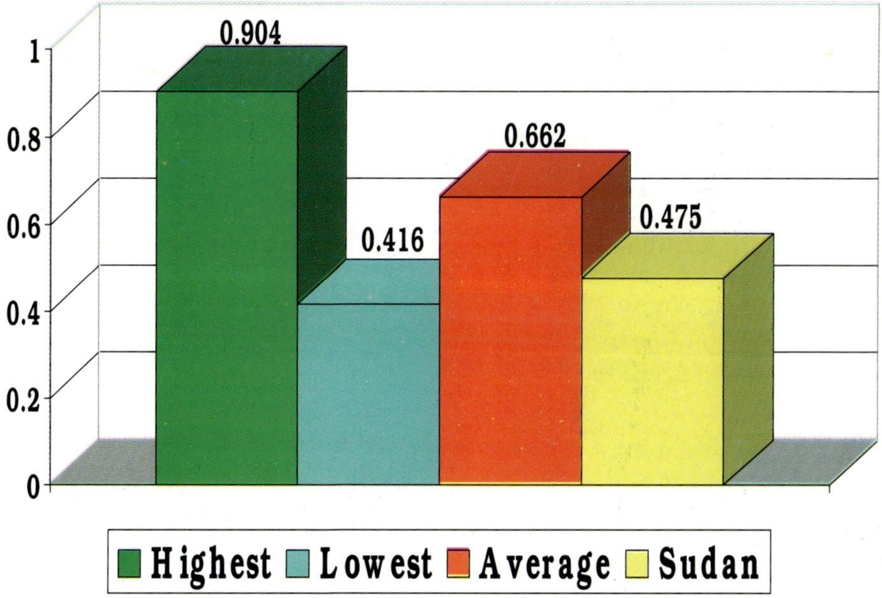

Figure 13.2: Per capita Gross Domestic Product (US$)

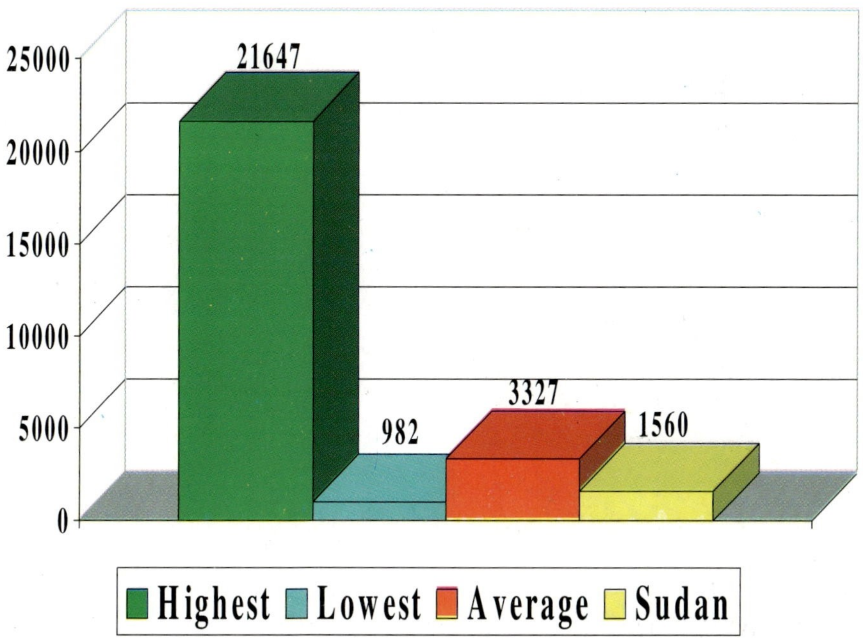

Figure 13.3: School Enrolment (School-University-Postgraduate)

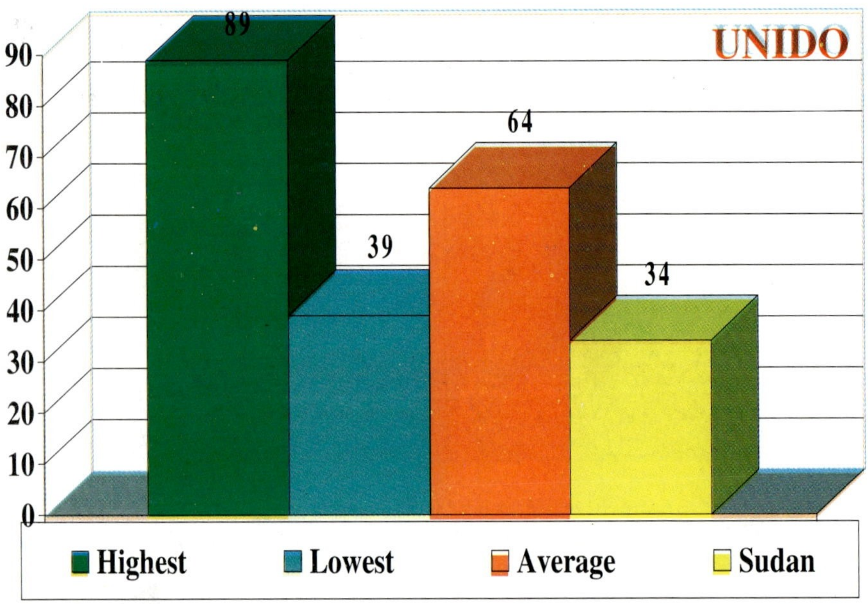

Figure 13.4: Number of State Universities

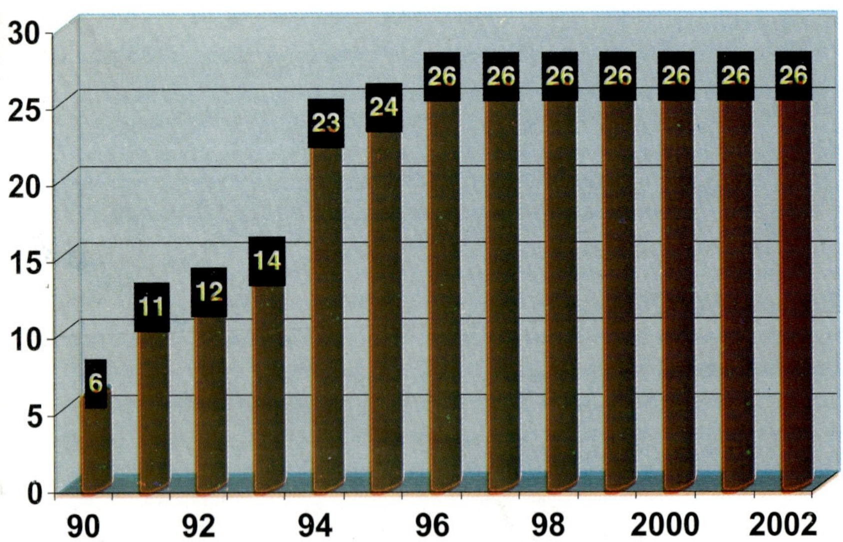

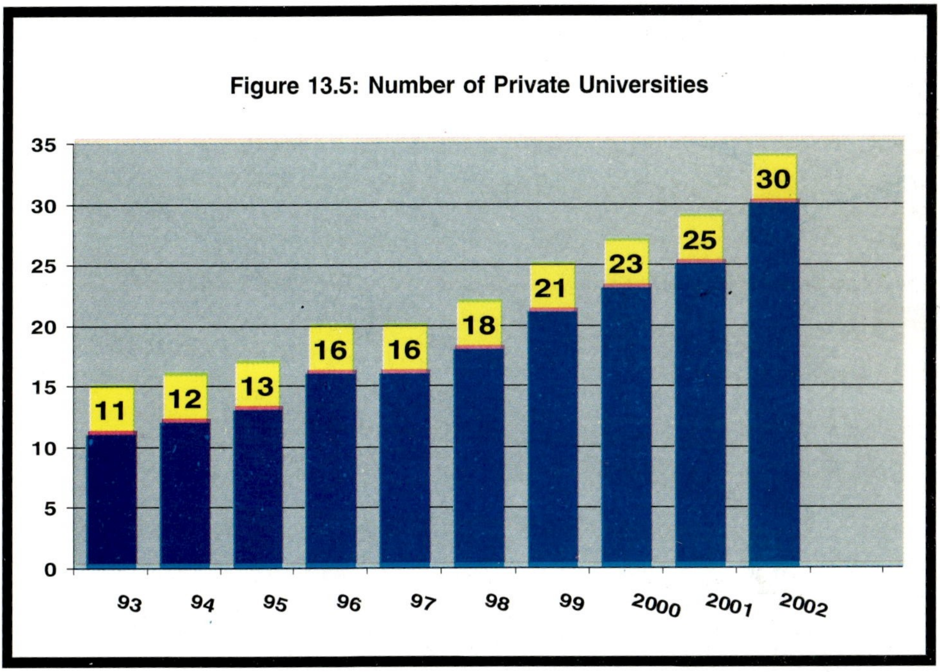

of the nation. The number of university graduates, scientists and researchers is on an increase due to the increasing number of universities, research institutions and mounting interest in postgraduate studies and scientific research. However, it is noticed that all these efforts are somewhat fragmented with duplication of work and irrational allocation of funds. It became evident that a qualified body is needed to co-ordinate these efforts, plan for science and technology research, development, and transfer, as well as finance, follow up the implementation of the national research plan and facilitate the dissemination and implementation of the tangible research results. To realize great expectations, the government established the Ministry of Science and Technology (MOST) in 2001. The mandate of MOST was to co-ordinate efforts, to create national and international links, and to formulate a national plan for science and technology. The main research centers and research institutions that were affiliated to the Ministry of Science and Technology were:

- Agricultural Research Corporation
- Animal Research Corporation
- The National Center for Research
- The Industrial Research and Consultancy Center
- Sudan Atomic Energy Commission
- Earth Sciences and Energy Corporation
- Economic and Social Studies Corporation

The sub sectors are directed to conduct research in agriculture, animal wealth, medicine, and to promote the national capacity in biotechnology research.

At the time when the Ministry of Science and Technology was established the units and staff of the ministry was as shown in Figure 13.6. The number of researchers per 1000 persons of the population was very low as seen in Figure 13.7. Also, the % expenditure in research to total Government expenditure was far behind other developed and developing countries (Figure 13.8). In general, the Investment in knowledge as % of the GDP for the country was small even for all the sectors of technology such as higher education, software, and Research and Development (Table 13.2).

Figure 13.6: Knowledge Investment as % of GDP

Sudan (0.002)

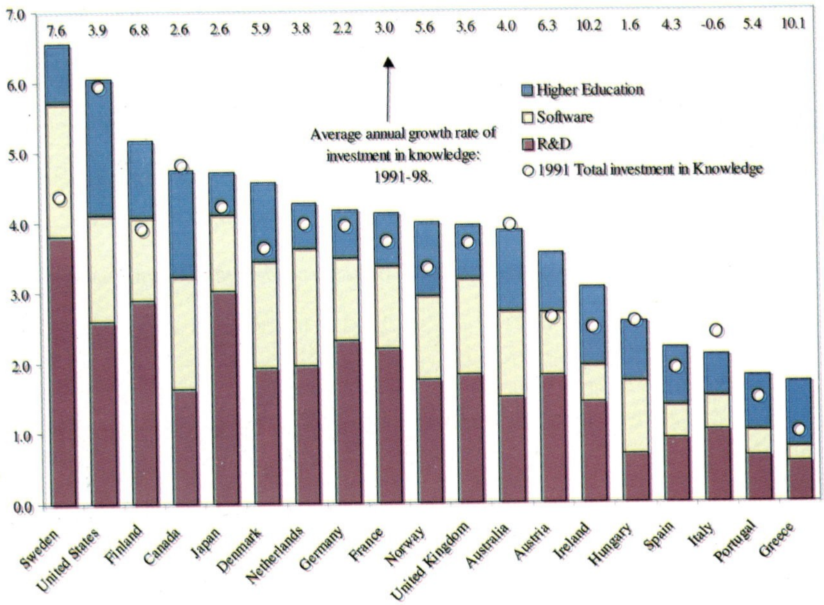

Table 13.2: Units and Staff of Science and Technology in the Ministry

Sl.No.	Institute	No. of Units	Human Resources			
			Researchers	Technicians	Labour	Total
1.	NCR	3 centres	279	355	173	807
2.	ARC	4 centres	423	1434	3455	5312
3.	ARRC	4 centres	282	548	608	1438
4.	IRCC	8 units	82	94	70	246
5.	NAEC	9 units	27	48	24	99
6.	ERI	4 units	22	18	30	70
7.	SERI	2 units	17	–	35	52
	Total	**28**	**1132**	**2497**	**4395**	**8024**

Figure 13.7: % Expenditure in Research to Total Government Expenditure

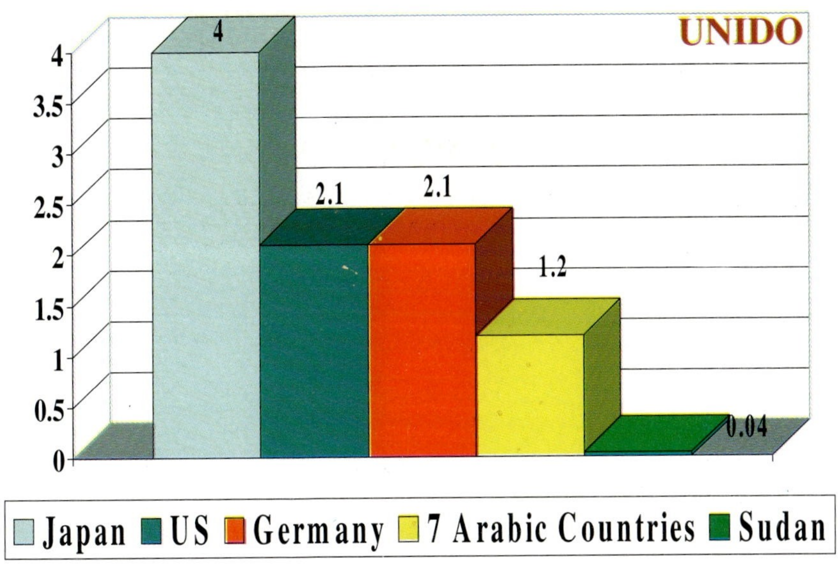

Figure 13.8: Number of Researchers per (1000) Population

Sudan (0.02)

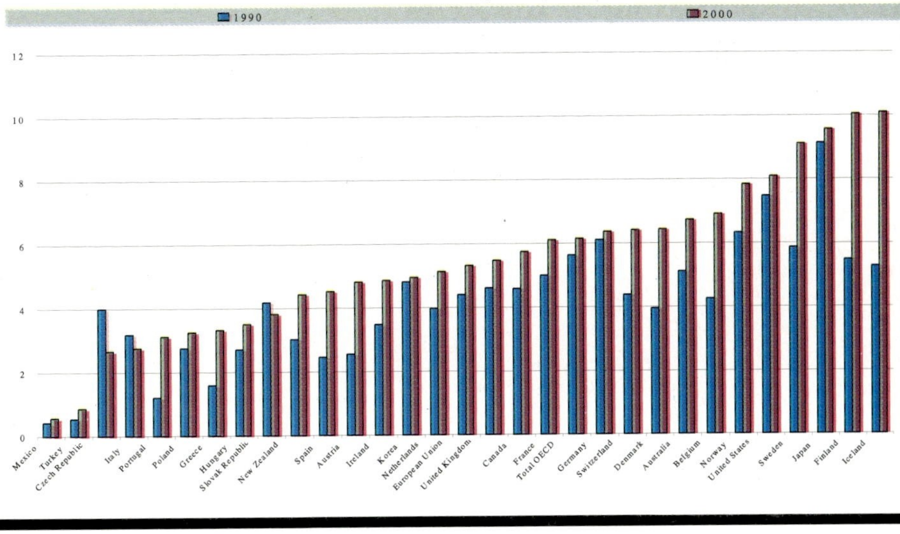

SUMMARY

To summarize, science centers and museums in Sudan had undergone drastic changes during the last century. The research centers and museums had expanded and covered most of the country's regions. It looks very clear from the analyzed data, that there is a big gap between Sudan and the rest of the world in the field of scientific research. It was evident that a major shake up in the country's policy regarding higher education and science and technology is eminent. The country is enjoying the first fruits of this policy adoption.

14

The Syrian Culture and Syrian Vision of the Role of Museums in Developing and Protecting the Culture and Civilization: The plans in the future

Yousef Khacho

Architect, General Director of Architecture and Museum, Damas Muflurin Shora

Phone: +963 93 709449, Fax:00 963 11 2452341/2247983

E-mail: ykhashow@ureach.com, antiquities@net.sy

Syria is one of the most important archaeological areas in the world because of its rich historical, archaeological and architectural heritage. Its ancient cities, castles and architectural sites are testimony to the historical centuries on the earth.

For a long time Syria was the destination of archaeological excavations and, as a result of all these great scientific excavations, we possess amazing quantities of archaeological finds and discoveries which have became symbols of this land's history. The idea of establishing museums began to clarify and specify the historical and cultural identity of this strategic region of the world. We began to establish the first museum in Damascus in 1936, which is considered to be one of the most important in the world because of its rich collections.

The material heritage is connected with the culture of humanity and it shows the progress of the civilization. By visiting a museum every citizen has the possibility to understand and "touch" the main stages of the human progress through the products of its material culture. Therefore, there exists a close connection and relationship between museums and human beings as the truthful expression of civilization. We can also consider museums as a means of communication between past and present and for this reason we must encourage men to protect and promote this valuable heritage by using the technology of the present time to add any modern methods and styles to protect this treasure and restore it.

The result of this operation is a shining future for all generations. It is important to understand that past plus present means future of the whole society. From this point the General Director of Antiquities and Museums in Syria intends to take care of the antiquities and the finds and build new museums in all parts of the Country. His aim is to give a complete picture about the ancient cultural and material heritage of a region to citizens.

The General Director is working on developing ways of explaining the archaeological results by using the new techniques that are required for exhibitions. What distinguishes our Museums is the variety of cultures and civilizations present in Syria.

The role of museums lies in affecting the culture of the nation and providing the present generation an idea about the historical depth of their civilization, and in searching for heritage and in developing ways to respond to the changing needs arising out of developments in a number of areas.

The museums are also of economic importance with great indirect benefits, compared to other economic activities.

Changing role of museums, in this case is leading to other functions, besides that of propagating the memory of happenings on this earth, as also in providing the country with greater number of tourists and in transmitting the shiny picture of the civilizations of this part of the world.

And the important point is the hard attack of the revolution of the globalization and the influence of the strong culture (technically, and informatics-wise).

Here will appear the role of the local culture and the history of the country which is depicted through the museums.

The General Director is working on restoration work, protecting the urban remains of the old cities and their architecture.

- Preparing scientific programmes to inform the people about their culture, and for external dissemination.
- Prepare museums for the memories of our cities connected to the traditional culture in the last two centuries to complete the picture of our civilizations to our citizens and our visitors.
- The Director is preparing many plans to develop museums all over Syria with all the activities, especially an expedition inside the museums and the castle, and the cities: I will show examples of the new ways of expedition which explain the story of the civilization using completely new techniques.
- Preparing a data base for museum archives.

15

Roles of the Monterey Bay Aquarium in Community and Marine Biological Education

Le Ly

Department of Oceanography, Naval Postgraduate School
Monterey, CA 93943 USA
E-mail: lely@nps.navy.mil; Website: www.oc.nps.navy.mil/~lely
Applied Science International (ASI) P.O. Box 7326 Silver Spring, MD 20907 USA
E-mail: lely_ASI@yahoo.com
Center for Regional Research & Development, Ministry of Science and Technology
70 Tran Hung Dao Str., Hanoi Vietnam
E-mail: lely_ASI@yahoo.com

ABSTRACT

This is my presentation on the role of the Monterey Bay Aquarium (MBA) in community and marine-biology education as I see it from my position of an MBA supporting member and an oceanic-atmospheric physicist. MBA is a well-known aquarium in California, U.S.A., and the world. It is owned by the MBA Foundation, which is a non-profit organization. It is located in Monterey County (around 200 kms South of San Francisco). The County has a very high level of education and many research and higher educational institutions. MBA opened October 20, 1984. Today it exhibits more than 600 species basically from the Monterey Bay (MOB) region. It has a giant 1-million gallon sea-water tank for tuna, shark, ocean sunfish and sea turtles. It is the first aquarium in the world to exhibit a living kelp forest. It has special sections, classrooms and labs for teaching.

MBA has many activities, involves more than 420 employees, about a thousand volunteers, and about 2 millions visitors per year. MBA is the number one aquarium and number three among all family attractions in the U.S. (It was rated so by Zagat Survey). It has a large annual budget made up from admission fees, membership, contributions from individuals and business, foundation grants, special events and gift and book stores. No government money is involved (except an occasional research grant). I would like to share my experience on MBA–I see it is a "non-formal learning institute"–that has an important role in community life and in research and marine biological education. I also have some studies on the Monterey Bay region to share in this paper.

INTRODUCTION

The Monterey Bay Aquarium (MBA) is one of the well-known aquariums in California, U.S.A., and the world. MBA is the number one aquarium and number three among all family attractions in the U.S. (It was rated so by Zagat Survey). It is located in the beautiful Monterey County (around 200-kms South of San Francisco). The County is famous with cities such as Monterey city, Camel city, with the homes of the world-class golf fields and beautiful weather around the year, and many famous international research and educational centres. It is one of the US regions with the highest numbers of Ph.Ds per square km.

In less than a 50-km distance from MBA, there are homes of famous international research and educational institutions such as Naval Postgraduate School (NPS), Naval Research Laboratory (NRL), Fleet Numerical Ocean-Atmosphere Center, Monterey Bay Aquarium Marine Research Institute (MBARI), Stanford's Hopkins Marine Station, National Ocean-Atmosphere Administration (NOAA), Pacific Grove Fishery Institute, Monterey Bay California State University (MOB CSU), and University of California at Santa Cruz (UCSC). There is a Monterey Bay (MOB) local Association of the above organizations, which focuses on joint projects and exchange research and operational information on physical, biological, and environmental studies of the MOB region.

MBA is a real "non-formal learning centre" and a science museum for Monterey County, Northern California, national, and international communities. The aquarium project was initiated in 1977 by four marine biologists at the Stanford's Hopkins marine station. The aquarium opened in October, 1984. In 1996 MBA has doubled exhibit space including the Outer Bay wing for the open ocean and deep sea. The main feature is the Outer Bay water exhibit, the one-million gallon (about 3,800,000 liters) giant tank for tuna, shark, ocean sunfish and sea turtles. The main view into this exhibit is the world's largest window of about 17 meters long and 5 meters high. Today MBA exhibits more than 550 species, basically from the MOB region. It was the first aquarium in the world to exhibit a living kelp forest.

MBA is owned by the MBA Foundation, a public non-profit organization established by Mr. Packard's family. The MBA was a gift to the community by the Packard family. The original cost of the MBA was approximately US$ 55 million. As a non-profit organization, the MBA's annual budget relies on income from admission fees, contributions and donations from individuals and businesses, foundation grants, special events, and gift and book stores. No government money is involved (except an occasional research grant).

MBA has active research and life-time learning programs. It has different groups working on sea otter conservation and tuna conservation biology. It also has various learning programs such as teacher & kids program, conservation & research program, and exhibits programs. MBA's sister institution, the Monterey Bay Aquarium Research Institute (MBARI) conducts deep-sea research in most of the Monterey Bay region.

ROLE OF MBA IN COMMUNITY, MARINE BIOLOGICAL RESEARCH AND EDUCATION

The mission and vision of MBA is to inspire conservation of the oceans.

MBA: A Life-Time Learning Centre and Its Role in Community

The MBA envisions a world in which the oceans are healthy and people are committed to protecting earth's natural systems. To help make this vision a reality, MBA creates exhibits, programs and publications that introduce people from all walks of life to the wonders of the marine world, from Monterey Bay to the vast oceans beyond. MBA offers enjoyable and enlightening experiences to inspire a love and understanding of nature. It is noted that effective conservation efforts have their roots in science; therefore MBA works to advance conservation through education and research. MBA helps people discover their connection with natural systems and become active ocean protectionists.

In less than a 50-km distance from MBA, there are homes of famous international research and educational institutions such as NPS, NRL, Fleet Numerical Ocean-Atmosphere Center, MBA, MBARI, Stanford's Hopkins Marine Station, NOAA, Pacific Grove Fishery Institute, and others with marine educational programs such as Monterey Peninsula Community College (MPC), MOB CSU, and UCSC. This condition enables MBA to have advantages in research and education, and close cooperation with neighboring institutions.

With more than 420 employees and about one thousand volunteers, and about two-million visitors per year with various activities and programs, MBA really is an important part of community life. MBA has many interesting programs that make it a public life-time learning centre for people of all ages. Next, some MBA typical sections with interesting programs are presented with more details.

Membership & Donors Section

This section has many programs such as "Membership rate & benefit", "Member events", "Planned giving", "Contact membership" and others and many of them can process member transactions online. The membership benefits include unlimited free admission (for a full year), no waiting–express entry through the MBA exclusive Member Entrance; special events (for member's guests also); 10 % discount on gift and book store merchandise.

"Member events" include many interesting happenings such as "Jazz at the Aquarium (June)", "Picnic by the Bay (July)", "Family Camp-ins" and others. Daily events at the aquarium include "Kelp forest feeding", "Sea otter feeding and training section", "Penguin feeding", "Outer Bay feeding program", "Exploring Monterey Canyon", "Surviving sharks" and many others.

Aquarium Exhibit Section

MBA has received many exhibit awards including "Jellies: Living Art" with American Zoo and Aquarium Association (AZAA) 2003, Significant Achievement Award and National Association for Museum Exhibit 2002-2003; "Excellence in Exhibition" Award; "Splash Zone" with AZAA 2002 Significant Achievement Award; "Vanishing Wildlife" with AZAA 2002 Munson Aquatic Conservation Exhibitory Award; "Outer Bay" with AZAA 1997 Best New Exhibit.

Currently, the aquarium has attractive exhibitions such as: "Kelp forest"; "Sea otter"; "Outer bay"; "Monterey Bay" with Rocky shore, Sandy beach and aviary; and Monterey Bay Deep Sea Canyon, and Marine mammals, and Octopus and kin.

Teacher and Kids Section

This section has various interesting programs and subprograms with many activities for teachers and kids to play with and learn from. Some activities, games and fun things to do such as "Coloring pages" (print personalized pages of favorite animals like sea otters, sharks and jellies); "Shark school of art" (learn some tips on how to draw sharks and create your own cartoons); "Make a tide pool game" (make a tide pool come to life in this popular online interactive. Learn about the animals that live in the rocky shore); "Be a sea searcher" (sharpen your observation skills! Search for numbers, letters, and patterns in the aquarium and at home), and many others have always involved kids. The aquarium has some classrooms and labs for teachers and kids programs. There is also the student oceanography club for students who are interested in marine science and conservation.

MBA ROLE IN CONSERVATION, RESEARCH AND EDUCATION

Marine Science Section

The aquarium cooperates with many other groups and institutions to understand and protect oceans. MBA has a direct close collaboration which includes programs: the Monterey Bay Aquarium Research Institute (MBARI) (the aquarium's sister institution focusing on deep-ocean exploration and research); Communication Partnership for Science and the Sea (COMPASS) (program mission is to accelerate the pace of solutions to important marine environmental problems); Monterey Bay National Marine Sanctuary (MBNMS) (federal protected marine area stretching over the central California coast); the ocean Project (focusing on helping people develop interest and caring attitude to take action for ocean conservation. MBA also has a nationwide collaboration among aquariums, zoos, and science, technology, and natural history museums to raise public awareness about the importance of the oceans to human survival.

There is also the Monterey Bay Association (MOBA) of the abovementioned local organizations, which focus on joint projects and exchange research and operational information on physical, biological, and environmental studies of the MOB region. The local association has some successes in exchanging research and operational information and joint projects on MOB region studies. It is noted that the MOB region has become a natural laboratory for scientists of different environmental disciplines such as environmental physicists, chemists, geologists, biologists, ecologists and others.

At NPS we have a large number of national and international multi-year, multi-organizational research projects and programs of studying oceanic and atmospheric physics of the MOB region, including state-of-the-art field observations, monitoring stations with satellites, radars, airplanes, and HPC forecasts and predictions of different time scales.

Have national, international multi-year, multi-organizational research projects on Monterey Bay circulation and tide with mathematical-computer state-of-the-art models, on air-sea interaction theory and modeling, and on a Physical-Biological Coupling study forward to develop a mathematical computer numerical model for Harmful Algae Blooms (HAB) prediction. The results of our studies have been circulated and reported in the local MOBA and published in national/international refereed professional-community journals (see also www.oc.nps.navy.mil/~lely). In particular, our research results had been recognized by national and international professional communities and we have been invited (among about 300 world-leading scientists in their fields) to write the paper: "Air-Sea Interaction: Surface Waves" for the Encyclopaedia of Atmospheric Sciences (first time in the world, for graduate teaching and research professionals in the field) published by Academic Press publisher (England), 2002, in a set of six volumes (3000 pages). We believe that results of our studies may make some contribution to national, international and local communities including the MBA.

SUMMARY

Overall, MBA is a good life-time learning institution for people of all ages. It has a very important role to play in improving science knowledge of communities, in ocean conservation, marine biology education, and research. To illustrate this, I have a short story. I have a colleague, who had a seven years old boy then. One day he happily visited MBA with his teacher and class. He came home from MBA and cried for fish dying from marine pollution and over oceans becoming empty because of over-fishing. The boy did not eat fish after that. He only ate fish again when he was over ten years old and knew more about the oceans and fish.

REFERENCES

http://www.mbayaq.org

http://www.oc.nps.navy.mil/~lely

Ly, L. N. and A. Yu. Benilov (2002). Air-Sea Interaction: Surface Waves. *Encyclopedia of Atmospheric Sciences*. J. Holton, J. Pyle and J. Curry (Editors). Academic Press, Vol. **1**, 118-127.

Ly, L. N., F. Aikman III, R. Stumpf, T. Gross (2004). Modeling of Cross-Shelf Transport in Support of Monitoring and Forecasting of Harmful Algae Blooms Along the West Florida Coast. *X-th International Conference on Harmful Algae. Proceedings* (in press).

Ly, L. N., W. P. O'Connor, J. D. Paduan (2000). A Barotropic Simulation of Semi-Diurnal and Diurnal Tides for the Monterey Bay Region. *Coastal Dynamics and Estuaries*, in *Hydraulic Engineering Software VIII*, W. R. Blain and C. A. Brebbia (ed.), Computational Mechanics Publications. 155-169.

16

More Effective and Attractive Activities are Needed for Museums to Universalize Scientific Knowledge

Nguyen Van Huy
Director, Vietnam Museum of Ethnology

The Vietnam Museum of Ethnology was founded in a socio-economic context at a time when the country had entered the renovation period for nearly a decade. The government's open economic policy was leading to economic changes, expansion of exchanges, application of scientific and technological achievements and the ever-improving standards of living for the people.

Formerly, in the period of government subsidies, the museums were not yet obliged to think much about how to "manage for survival" by themselves. The activities or products by museums were not offered on the basis of taking into account the question of how these products would be accepted by the communities, or, in other words, the objective of "serving the people" was not yet seriously paid much attention to. Museum shows were mostly made for the purpose of propaganda, and were completely subsidized. Government subsidies were gradually leading to stagnation in museums, making creative and imaginative museum activities fade. However, today's museums are no longer standing outside and away from the laws of the market economy. It is impossible for museums not to concentrate on the subjects of their services and their clients; it is now a must for the museums to mobilize and find ways to attract the masses to visit the museums, for it is the masses that bring an important source of revenue for the museums. Thus, it is due to the objective demand (the demand for cultural appreciation from the masses) and the subjective need of the museums (the need for increasing revenues to sustain professional activities) that they have to diversify their activities with an ever-increasing quality.

As it is mentioned above, the masses play a vital role in the survival of museums; therefore, it is extremely necessary to investigate the needs of the masses. Today's masses are not likely to easily accept any kind of "food for thought" offered by museums, for they have their own preferences. Acting as an institution which studies and propagates for the life and civilization of 54 nationalities currently present in Vietnam, the Vietnam Museum of Ethnology turns its attention to complicated and multiform masses as clients, including both Vietnamese and non-Vietnamese citizens, people from the capital city and provincial areas, people of different ages, different sex and different cultural levels. Among these, people from the capital city, especially children (school pupils) and foreign visitors, are paid special attention by the Museum.

Being a national museum located in Hanoi, the capital city, the Vietnam Museum of Ethnology determines its task as attracting and providing services for the capital visitors in the most effective way. Hanoi, as the capital city of the country, is a place where museums are concentrated, including among them national museums and sectoral museums; but the ground reality is that the masses are still indifferent to museums. Many city-dwellers in Hanoi prefer to go to market-fares or localities away from the city instead of going to museums. Besides, museums have to compete with other entertainment centers of the city such as theatres, movie houses, artistic performance centers, water-parks etc. This is a challenge to museums in general, and to the Vietnam Museum of Ethnology in particular. Why is it necessary to attract people from the capital to the museums? According to experience of other museums in the world, a museum can "live" only by attracting a large number of visitors as its clients. They are mainly the local residents in the place where the museum is located. They are the first people who accept, and are influenced by, the activities of the museum. Furthermore, they are the people, with distance advantage, given more favourable conditions to visit the museum more often. In addition, residents here also have a great demand for knowledge and cultural appreciation. As a result, the capital city residents are really a class of clients that museums should turn special attention to. Apart from that, museums should manage to find an answer to the question of reaching people all over the country, those who have not been able to travel to the capital city of Hanoi. How can the broad masses of people, especially the young living far away from Hanoi be helped to access the contents as expressed in the shows by city museums?

Additionally, the fact that the number of international visitors to Hanoi is always on a rise is a favourable condition for the museums. The demand for visiting museums by tourists from European, American, or other Asian countries is quite large, for visiting museums has become a habit of tourists. In particular, tourists who wish to learn about the habits and customs of the country and of the people of Vietnam often put a high priority on visiting the Vietnam Museum of Ethnology before visiting other places. In addition to that, being located on the highway connecting the Noibai International Airport and capital Hanoi, the Vietnam Museum of Ethnology is a convenient place to stop by for international tourists before arriving in Hanoi or leaving for the airport. That advantage in tourist potentiality enables the Museum to bring into full play its role as an institution which introduces the cultures of different ethnic groups in Vietnam to the people of the world, and at the same time, increases its revenue for organizing specialized activities.

In order to attract the masses to the museums, the museums should first and foremost understand the needs of the people, on the basis of which, to create activities which meet these needs. It means that the museums should not organize their shows and activities on a subjective basis without taking into account the interests, the needs and the expectations of the visitors. Much thinking and innovation should be invested in the shows that "serve the anniversarial purposes" but do not take into consideration the needs of the broad masses; and in the shows that only display what is available with the museums and what museums want to communicate or propagate to the visitors. Museums now should no longer wait for the visitors but attract the broad masses to the museums, through a variety of activities which suit the needs of the broad masses. Thus, only with multiform shows and displays regularly organized with great attractions and great educational effect, plus a great volume of information to the masses, museums in general can "reach out" to their visitors. People are willing to go to museums for their learning, enhancement of knowledge and experience; and achieving this needs time and really is a process of hard striving for all the museums.

Starting with the conception that museums must reach out to the masses, ways to organize activities of the museums were modified as compared to what was done in the past years. As mentioned above, formerly, museums were also interested in specialized activities such as research, collection, conservation, display, education etc. but the importance of the masses and the communities was not really appreciated. If, in the previous years, research was only carried out at communities for information collection, now, it is carried out with more emphasis on cooperation, and collaboration from the communities in order to conserve and bring into full play the value of the cultural relics. The majority of displays done in the earlier period were inclined to serve the purpose of purely simple propaganda, and as a consequence, they were unattractive and even boring. Educational programmes suited to particular or specific classes of visitors, especially school children, were lacking. Particularly, the simple conception holding that display was the central activity of a museum was widespread, and consequently after being opened, a museum display was regarded as having been completed, without any awareness of having to organize multiform activities to attract visitors to museums from the opening to the closing of the display. These are artistic shows and performances, workshops, seminars, film shows and many other types of activities of the educational programme. Nowadays, activities that accompany museum displays have become frequent and indispensable at the Vietnam Museum of Ethnology. Following are some specific examples:

TOPICAL DISPLAYS

Topical display plays a very important role in every museum, constituting one of the key activities of a museum. If permanent display can present/ raise a problem in a general way with some emphasis on certain points, a topical/ temporary display is an opportunity for a museum to exploit a certain aspect of specialization which a permanent display is unable to present. For example, in a permanent display at the Vietnam Museum of Ethnology, the section that introduces the Laos is given at a glimpse, in an extremely brief account, but in the topical display on "Handicraftsmen and Photographers", the museum clearly and distinctively figures out the life and traditional weaving trade of the Laos at the commune of *Nasang 2* in *Laichau* province. Furthermore, visitors to the museum can have opportunities to meet and get into contact with original weavers from the *Nasang* commune, in two days after the opening of the display. In addition to this, other activities such as workshops, performances etc. are also organized. Thus, topical displays such as display of the traditional weaving trade in *Nasang* 2, of the Vietnamese traditional medicines in *Daiyen* Hamlet, of the bamboo-weaving trade in the *Khmu* ethnic group etc., are a driving force for visitors to come and visit the museum, creating new, regularly organized, attractive activities which are held for visitors not only once, but several times. Consequently, it is necessary for the museum to make plans to regularly organize topical displays, for it is the topical displays along with other activities that accompany them, that will create an attraction for the broad masses to come to the museum.

The current trend is to encourage cultural subjects to participate in museum activities, to stimulate them to present their own cultures and their own thoughts (photovoice), to make community-linked and development-linked research, (conservation and development of traditional handicrafts) popular arts such as the Tay ethnic group's puppet dance, and water puppet dance. Nowadays, museums are no longer a place where only the museum's research results on the cultures of different ethnic groups are displayed, but a forum where the cultural subjects can present themselves, introducing themselves to the broad public. This enables the museum visitors to enjoy the opportunity to meet cultural subjects in person while new invigorating and interesting experience may be created for them by the museum.

The diversification and extension of space for topical displays can at the same time open an approach to other types of displays. But what can be done to make topical displays attractive? Attractiveness is by itself dependent on a number of factors such as display techniques, lighting, colouring, content etc. However, there still remains one very important unanswered question for each and every display: what is the starting point for a display? A collection of objects available at the museum which may be shown to the public or the needs perceived from the public, the society? The latest worldwide displays in museums are inclined to follow a tendency, that of satisfying the needs of the museum visitors, of the general public. That a visitor goes to a museum derives from a simple fact that he or she is interested in seeing something which they need to see there. As a result, it is obligatory for a museum to investigate, or do research on the needs of the general public before making a display, so as to make a prediction on what messages would satisfy the needs for knowledge of the public. Displays should be closely associated with contemporary needs of the society. Only in such a way could displays attract visitors. The more visitors attracted, the greater the cultural, the social effect will be and with these there is an economic benefit for the museum. It is quite clear that economics (between expenses and revenues) for each display should be worked out. It is therefore impossible to neglect the needs of the public if museums want to organize displays which satisfy the needs for dissemination of knowledge on the part of the organizer and the needs for knowledge on the part of the museum visitors.

In order to be aware of the needs of the public, it is a must to emphasize the importance of evaluating the behaviours of visitors in all phases of the display: pre-display, during-display and post-display. It is the opinions of the visitors that will perfect the arrangement of the display.

EDUCATIONAL PROGRAMME

Today's museums should have a well-founded awareness of the role played by the educational programme in museum activities. It is not a propaganda programme but a highly specialized programme of pedagogical nature. It is an indispensable activity of a modern museum.

Educational programmes in museums should pay attention to the young people, especially school children. Education by means of museums is called informal education, which means it is not in the form of "formal education" as provided by schools where knowledge is imparted to students by teachers who train by giving official lessons in class for their students to acquire knowledge. Education by means of museums is a programme which creates conditions for learners to discover new knowledge by themselves; learners are allowed to learn in an "informal" environment; therefore, they are under no pressure, but are encouraged to learn according to their free will. In addition, museums also provide teaching aids for students in the form of objects which are tangible, which they can feel directly by their senses... with the result that they can perceive with the smallest details the things they are interested in. Such an educational environment as the programme which teaches how to make earthenware, or how to use the Vietnamese traditional medicinal plants and trees or how to play the folk games as provided by the Vietnam Museum of Ethnology, is an example of how to combine understandings of school children's background knowledge, psychology, and needs with the aim of disseminating and conserving knowledge of traditions set forth by the museum. Such programmes are given wide support by school children, as well as by handicraftsmen or professional artists and the society at large. This illustrates the correct approach as used by the Vietnam Museum of Ethnology with regard to the issue of how to handle the educational programme.

The Museum is also interested in the quality of the visits. The Museum has a different viewpoint on how to respond to school children who visit the Museum, apparent in the Museum's plan to upgrade inventiveness, creativity in school children, in organizing attractive, vivid, suitable activities which fit the psychology, health conditions and expectations of school children, by allowing them to participate, to touch, to discover, to operate, and at the same time, to supplement their knowledge before, during and after their visit to the Museum. (To be able to do this, it is necessary to have a close relationship with the school, organizations of the youth union, young pioneer (adolescent) union, and at the same time, to investigate the content of the school textbooks, school curricula with a view to consolidating the Museum's understanding of the subjects served). The young people of today are especially interested in application of technological advances. In order to make the Museum an attractive place for them, it is essential to introduce technological advancements into the Museum, especially those in electronics, computer science and digital technology.

Based on the abovementioned viewpoint and on the new awareness of museum education, the Vietnam Museum of Ethnology has been putting together appropriate activities which fit the different types of museum users as follows:

- Constructing client-oriented sightseeing tours for schools and school children, with learning materials designed for school children of different age groups, and with instruction materials for teachers. During the recent three days from September 26th to September 29th, the Vietnam Museum of Ethnology opened the 4th training workshop for teachers from different schools in Hanoi on how to introduce the values of cultural relics into schools and how to effectively exploit real objects available at the Museum, as a follow-up activity after the workshops held at Angiang province, HochiMinh city, Cantho Province. Programmes like these will be continued.
- Designing sight-seeing tours for families, and instruction materials for parents as a guide on how to make a visit enjoyable at the Vietnam Museum of Ethnology.
- Designing a variety of educational activities such as activities that may be joined in at permanent or topical displays by the Museum, building a discovery section and on-the-spot forms of discovery with mobile carriers suited to each kind of display chamber, designing an educational programme on cultural relics such as giving instruction on traditional handicrafts for school children, on making earthenware, printing Dongho paintings, playing folk music and singing folksongs.
- Setting up and expanding applications of new technologies, especially information technology, electronics technology in the museum activities such as construction of websites in which attractive games associated with the objects demonstrated at the Museum are available, construction of a website museum where scientific discovery programmes such as Try Science by IBM are being applied; these are the programmes which the young people are especially interested in. Children may get access to these programmes on the webpage www.TryScience.org; this is considered to be the first worldwide virtual scientific and technological center. Through this webpage, school children, parents as well as teachers may get access to and experience scientific fascinations from different museums in the world through interactions with online adventures. Getting access to TryScience kiosk, children may take part in discovery activities through "tasks" as provided by scientific centers all over the world. Children may select one of the following activities: archeologist, space explorer, whale, and challenge.

"Archaeologist"

The child taking part in the game becomes an archeologist seeking out fossilized dinosaurs. The game players have to discover and identify the types of fossils before they are drawn away by running torrents. The game players may learn something about the process of archaeological excavation for fossils by taking part in a life-like one and by using specimens of fossilized patterns to compare and identify the discovered fossils.

"Space Explorer"

The task is to save the international space station (ISS). Suppose the ISS were struck by an unexpected incident and the game players were requested to find ways to avoid the danger that may be caused by the decreasing pressure in the spaceship! In this activity, the game players will learn how to become an astronaut on the ISS, and learn the way to move the cargo carrier from the shuttle spaceship to fit it in the ISS, the way to discover the resident chamber of the astronaut and the way to get out from the ISS to make necessary repairs.

"Wildlife in the Seas: Discover it with the Whale"

In this activity, the game players will help the whale to find food, to avoid dangerous carnivorous animals and to look for its shoal. In the process of the game, the game players will learn more about the animals living in the seas, about the way whales locate their route by using ultrasonic resonance at night, in the dark waters and the way the whales communicate with each other.

"Conquest of the Racetrack: A Challenge"

The game players will take part in a hard race with ordeals created for them aiming to help them learn more about physical science. The game player will have to compete with other game players getting access to the science museum on the net from different stations. As a consequence, the game players have to be on a quick response, otherwise their competitors from other stations in the world will win! To achieve success on the racetrack, the game players have to overcome five obstacles. These ordeals will teach or test the player's knowledge on piercing movement, theory of balance, optics, response time and muscle movements.

In addition to the science museum on the net, TryScience also contains other clusters of activities.

Constructing computer assisted learning programmes aiding the learners to discover various cultural relics.

Programmes for the General Public

Alongside the programmes designed for learners of different age groups, the Museum also designs a number of other programmes intended for the general public.

- Concentrating on construction and formulation of performance and presentation programmes associated with displays on the theme of culture and the community. The Museum invites members from different ethnic communities to introduce or make presentations on their own culture. The Museum has art-performance programmes for specific purposes, such as the water puppet programmes which have been organized on a monthly basis for about 2 years now and workshops for presentation of handicrafts production skills accompanied by demonstrated objects. These are activities which have attracted much attention of visitors.

- The audio-visual programmes. Audio-visual aids (videos, music) play an important role in the activities of modern museums, especially in display activity in general. They contribute to the attractiveness and vividness of the demonstrated objects and make it easier for isitors to appreciate the display. It is necessary for the Museum to organize audio-visual programmes in close relation to the contents of the display (by creating new films or researching to find films which have contents associated with the topic of concern).
- Organizing lectures, workshops, round table talks which are closely related in content to the topical displays.

As such, with varied activities and creativity, the Museum has played an active role in finding out about solutions for satisfying the needs for cultural appreciation of the general public, and for enhancing the knowledge level of the public, with an ever-increasing quality and as a result, the Museum has been successful in maintaining the number of visits frequently made to the Museum and in getting access to new visitor population from the general public.

CONCLUSION

Diversification of museum activities has to serve public with an ever-increasing quality of service.

As a first prerequisite, a new concept is needed on how a modern museum should be like, of which the important issue is that the museum should satisfy the needs of the contemporary general public, the museum should take part in the competition with innumerable other social activities to attract the general public to its services.

As a follow-up step, the museum should set forth an objective for its service that is to turn to the general public, with a service designed for the public.

It is necessary for a museum to be deeply aware that the public is diverse and multiform in terms of their needs which the museum should meet and satisfy. Special attention should be given to young visitors, by museums.

It is also necessary for a museum to be deeply aware that renovation of museum activities, along with diversification of these activities should be carried out in a public-centered approach.

Museum officials should be highly qualified, highly specialized so as to satisfy the demand for community-oriented service.

17

Vietnam Strives for Purpose to Socializing Science and Technology Activities

Phan Huy Chi
Director
Centre for Regional Research and Development
Ministry of Sciences and Technology, Vietnam

In recent years, the words *"socialize the science and technology activities"* have been mentioned in many conferences, reports and communication meets in Vietnam. The Vietnam Goverment is more and more concerned about it and has taken many decisions to develop such activities, through technical competitions, strengthening of museum activities, programmes to apply science and technology to developing social ecconomy and bringing science and technology information to remote areas.

In fact, efforts to "socialize the science and technology activities" has been remarkably stepped up, and the conference being hosted by Vietnam today is also a demonstration of the same intent.

To *"socialize the science and technology activities"* for the developing countries is of important significance. We are concerned about the aspect of human resources training and universalization of science and technology knowledge for the community in this Conference. In this age of science and technology explosion, to spread awareness about science and technology and to improve the scientific literacy of the community is very necessary to shorten the gap between the rich and the poor countries. There are various ways to approach these objectives:

- Set up training courses, mechanisms for informal education, seminars, conferences etc.
- Use all possible communication means.
- Publish documents, magazines, leaflets on a variety of subjects of interest for dissemination.
- Organize entertainment activities, skill competitions, technical skill competitions, knowledge competitions etc.
- Establish science centres, scientific museums etc.
- And more...

The Science Centres and Scientific Museums can play a big role here, and it is very important to make use of them to universalize knowlegde for all the communities, and to create a passion for learning to make our life better and happier.

There are many ways to look at the role of Science Centres and Scientific Museums, but the most effective is: A Science Centre or a Scientific Museum is the place, where people, in every age group, are able to learn according to their own abilities; this is the place which could satisfy the curious mind and encourage spreading of knowledge by using appropriate skills and methods to present natural phenomena and happenings around in our day-to-day lives and their scientific explanations through questions and anwers. This can be made enjoyable and interesting for people and this will make the Science Centres and Scientific Museums socialize more and more strongly and significantly. It is something new for Vietnam to set up Science Centres for universalization of knowledge through self study and self improvement. It requires certain skills to design and set up such centres and it is neccesary to research and use creativity to get better results.

In some aspects, Vietnam is not as developed as many other nations in the task of organising Science Centres and Scientific Museums in a way that they play their role correctly. For this we had done a study and a survey in some neighbouring countries. But generally speaking to universalize science and technology activities to improve the awareness and knowledge level of the Vietnamese people, we would need to use a wide variety of means such as television programmes, Robot-cum creative competitions, Vietnam intellectual contests and programmes aimed at bringing science and technology achievements, science and technology information to the moutainous and rural areas etc. The Museums' activities have been developing strongly in Vietnam, especially the activities of the Vietnam Ethnological Museum, even though it is a new Museum, it has had a good impact.

Presently, there are some 115 museums in Vietnam, but their major role has been to preserve cultural heritage, and historical objects. There are not many specialized scientific museums, particularly for study, entertainment and dissemination of knowledge. There are fewer places for deploying creative scientific ideas etc. on how to organise a museum in which attracting visitors is more important. We hope that through this Conference we shall learn much from all of you

We do hope that through such conferences we get an opportunity to learn many good things from you in respect of organising museum activities which are more and more effective and useful. At present, the most desirable thing for us is: how to attract visitors to Science Museums and Scientific Centres. The obvious answer is that we should raise their attractiveness by making use of better methods of organising activities and form of presentation and content. The experiences of foreign countries is very precious to the Vietnam's Government.

Today we had an interesting development, when a Malaysian Company learnt that the NAM S&T Centre is joining hands with Vietnam in organising this conference with the specific objective of introducing a model of science and technology centre to popularise the basic scientific knowledge among the communities and this kind of a centre would be set up in Vietnam, they came to talk to us. We highly appriciate your goodwill and wish to learn much more about this subject from you all.

We warmly welcome the participation of all the ladies and gentlemen who have come from many developing and developed countries. We welcome all the delegates who are concerned about the content of this conference today; all the reports presented by the participants have been very interesting and full of information that would help open the way for Vietnam to move on to the path of "socializing the science and technology activities by an effective way of organising the Science Centres and Science Museums".

We do hope that after the conclusion of this conference, all the participants would reach agreement on terms for supporting setting up of Science Centres and Scientific Museums, and that different countries would be able to operate these institutions for the role , intent and purposes that they are expected to take on. We also hope that a department headed by a regional secretary, will be established in Vietnam to coordinate with other nations and support the setting up of and developing Science Centres. We also expect that we would be able to demonstrate "the changing role of the Science Centres and Scientific Museums in developing countries".

Thanks for your interest!

18

Zambia's Advance Towards Developing A Science Centre

Festers Hangandu Mungo
Principal Education Officer, National Science Centre, Zambia
Phone:+260 1 263391, 266772 (O), 250571(R), Fax: +260 1 263391,
E-mail: nsc@zamtel.zm, dsikanyiti@yahoo.co.uk

INTRODUCTION

Zambia is a landlocked African country situated in Southern Africa. It covers a total area of 752,614 km^2 with a population of twelve and half million comprising seventy three tribes.

Figure 18.1: Map of Zambia Showing the Land Locking

There are four large rivers and four lakes in Zambia [Figure 18.1]. The rivers are: the Zambezi, Kafue, Luangwa and Luapula, shared with the Democratic Republic of Congo (Congo DR). The lakes are Mweru, Tanganyika (second deepest in the world) and shared with Tanzania and Congo DR; Kariba (is man made and) shared with Zimbabwe, and Bangweulu shared with Congo DR.

Copper is Zambia's main export earner accounting for 80% of the total exports. Other minerals are Cobalt, Lead, Zinc, Coal and several types of gemstones.

The present government has laid stress on the need to shift from the dependency on copper to agriculture and tourism among others. This is already showing positive results.

Since independence (1964) Zambia has tried to follow various macro and micro development strategies, most of which worked fairly well up to the early 1970s. However, between 1970 and 1980, Zambia experienced stagnation in economic and industrial performance.

It is in this regard that my government has made great efforts to promote science and technology through various fora such as Junior Engineers, Technicians and Scientists (JETS) and National Institute for Scientific and Industrial Research (NISIR). JETS activities are conducted in schools where school boys and girls are encouraged to make scientific projects/models for display at science fairs every year throughout the country. Thereafter a national and major science fair is held at the University of Zambia annually.

The science activities that school children conduct do present a "thrill" of science to them as a supplement to formal school education and to some extent this manifests itself into careers in science. It generally gives children the opportunity to experience the flavour of research and development. As a Centre we do provide support in the form of technical advice and sometimes materials when requested and if available in our stock.

THE ROLE OF NATIONAL SCIENCE CENTRE

Zambia has only one science centre called the National Science Centre (NSC). The NSC was established in 1991 with a major core business of "**upgrading of science and mathematics teaching and local production of science equipment".** NSC has two support centres referred to as *Satellite Centres* situated in two provinces, one in the North and the other in the South of the country. Both the National Science Centre and its Satellite Centres are production and delivery points. NSC operational objectives include:

- The development of a capacity for the local production of science and technology teaching equipments and materials.
    - ❖ The NSC has designed a Mobile Laboratory meant for use even in schools in the remotest areas (Figure 18.2).
- The planning and initiation of an in-service programme for science and technology teachers and facilitation of the production of low-cost teaching materials [Figure 18.7(b) and Appendix].
    - ❖ Teachers of science and technology benefit through workshops and seminars on matters of improvisation and development of learning

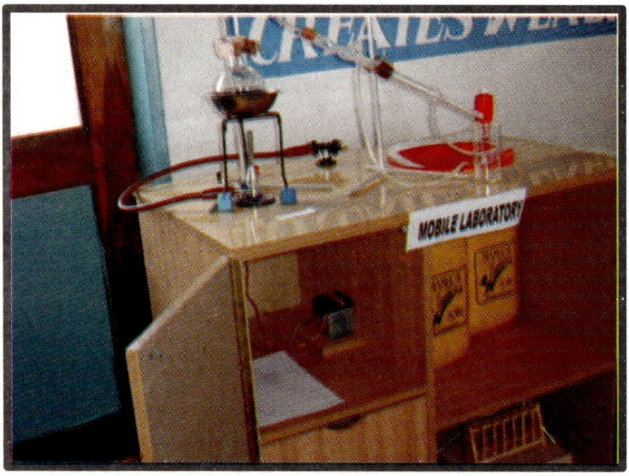

Figure 18.2

materials. During such events teachers are shown how to improvise teaching aids and also how to make low-cost teaching materials using, as much as possible, locally available resources. Generally, this enables teachers to enhance their teaching skills of science and technology subjects.

The reason why the NSC was made to focus on the promotion of science and technology specifically was because of the zeal of the Zambian government in wanting to move with the world in those areas. This is why they are even able to send a representative to this important gathering.

INTERACTIVE LEARNING

The Zambian National Science Centre can be said to be fairly progressive in the sense that since 2002, with the Kolkata exposure, there have been some positive changes.

Figure 18.3, for example, shows new developments at the Centre. The Centre is currently developing a Science Park, which is expected to have eight science exhibits initially, with intentions of expanding it in due course as more ideas become available.

So far our experience has shown that a Science Centre being a "physical venue where interactive exhibits and activities are used for imparting non-formal science and technology education to a general audience", we should try to convert this thinking into action.

We, therefore, wish to see a situation where our existing centres will some day:

- Popularise science generally
- Fight superstition and obscurantism
- Organise outreach programmes and generate a scientific culture within the society

SHOW EXHIBITS

The NSC has been represented at various conferences both locally and internationally. On the local scene we have attended the International Trade Fair and the Agricultural and Commercial shows where we exhibited our products [Figures 18.4, 18.5, 18.6(a), (b) and 18.7].

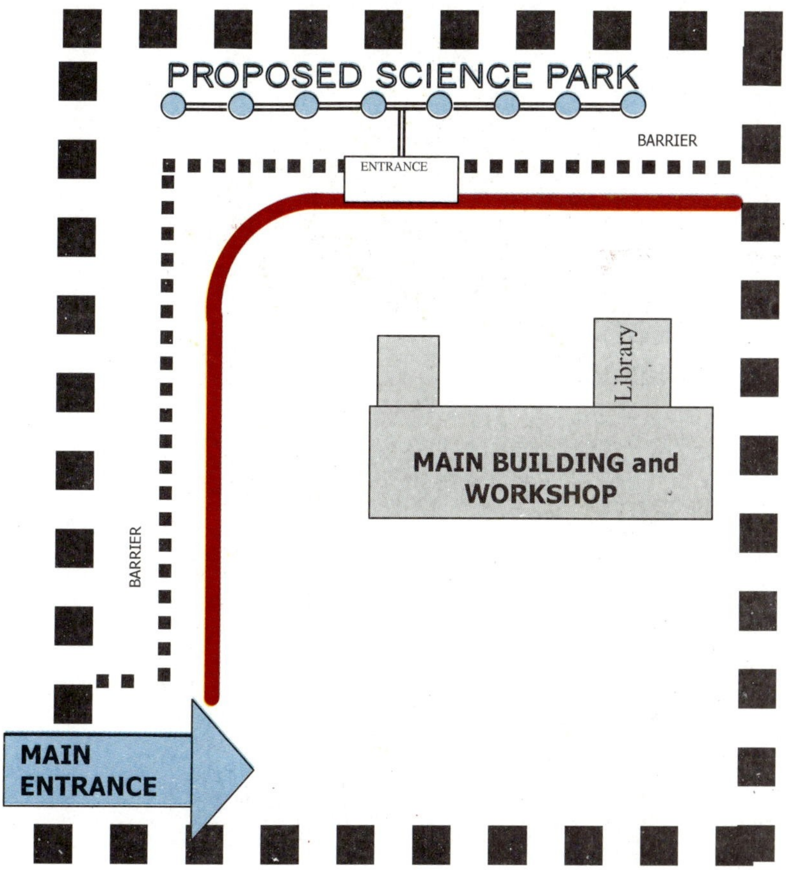

Figure 18.3: The Plan of the Science Centre Showing the Proposed Science Park

Figure 18.4 shows music tubes, an idea learned from India's Bardhaman Science Centre during the workshop on "The Role Of Science Centres In Development Strategies" held in Kolkata, India, between 30th November and 2nd December, 2002.

This particular exhibit was made specifically for the Zambia Annual Commercial and Agricultural Show held in Lusaka, Zambia, from 29th July to 3rd August, 2004.

A total of 3,013 show goers visited our stand. The theme of this year's show was "PRODUCTION CREATES WEALTH" (Figure 18.5). As a Science Centre our dream

Figure 18.4

Figure 18.5

is to be part of Zambia's wealth creating process through the promotion of science and technology.

INTERNATIONAL COOPERATING PARTNERS

Since 2002 the Centre has had technical assistance from the Government of Japan through senior volunteers who are attached to the institution as advisors on contract of two years each under the auspices of the Japan International Cooperation Agency (JICA). The volunteers have been contracted to help the Centre in the design, development and improvisation of learning materials. They have also been participating in the commercial shows [Figure 18.6(b)].

ZAMBIA'S NATIONAL EDUCATION CHALLENGE

The Government of the Republic of Zambia, in its recognition of the value of science and technology in schools, has stated in the Education Policy (1996) that "Science and technology have had a radical impact on Zambia's economy and on the

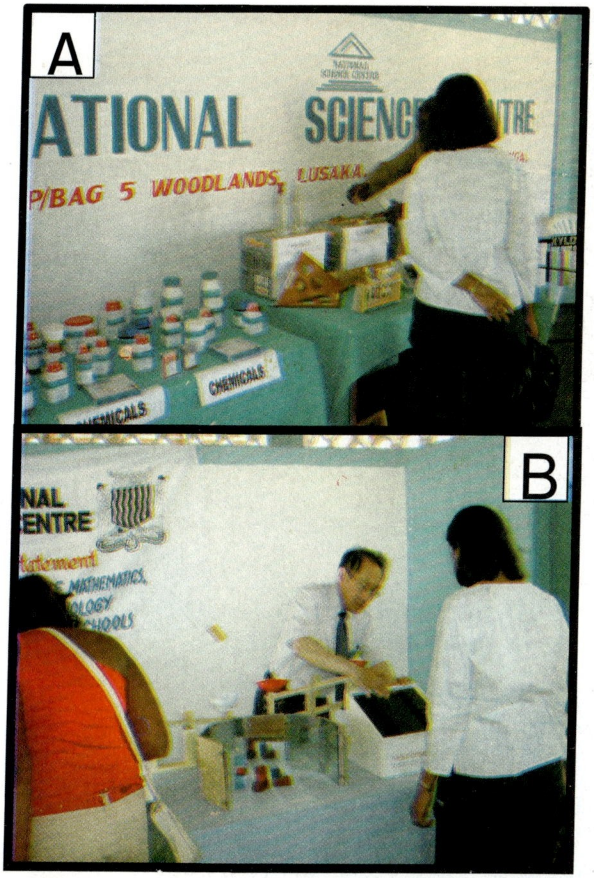

Figure 18.6: NSC Exhibiting at the 2004 Agricultural and Commercial Show

way of life of almost every one of her people" (pp 35). The document further recognises that "The ability to think scientifically and to understand scientific processes is becoming a condition for survival". This approach strengthens the idea of fighting superstition and obscurantism through the promotion of science.

As a result of this stand it is still the position of our government to provide sound science knowledge through the provision of support in every way possible as it will not be sufficient to teach science in theory alone. It is anticipated that with the reinforcement of the existing Centres in the country pupils will not only be told about certain processes, or even that they can observe them only, but that they should apply their own ideas, use their own hands, conduct their own investigations, however simple.

School children visiting the NSC stand during the 2004 commercial show.

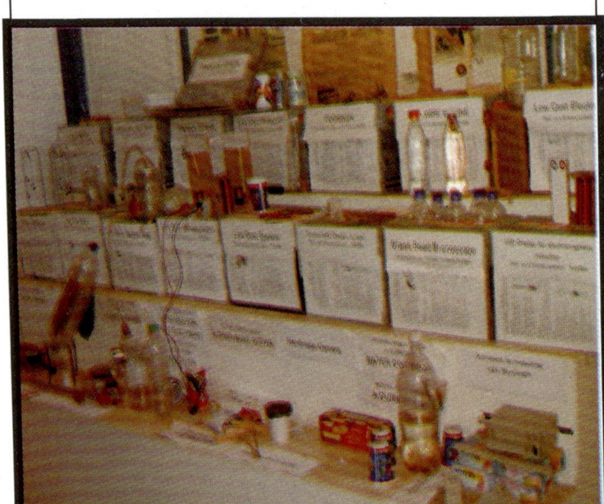

E 7

A display of improvised low-cost teaching aids.

Figure 18.7(a) and (b)

CONCLUSION

The need for the knowledge of science can not be over emphasised. However, it is up to those with the knowledge to learn to share that knowledge with the "underprivileged".

Those with the interest should not be discouraged by those with the knowledge. For example, countries that are already ahead of others in the development of science centres should be able to offer cost effective proposals especially for less developed countries.

APPENDIX

Listed below are some of the devices that have been improvised for schools as a cost effective way of providing the necessary teaching materials to schools

CONDUCTIVITY CHECKER

This apparatus can be used to check conductivity of electricity and also to test a circuit. If a substance is a good conductor, the LED on the left end of the checker lights up.

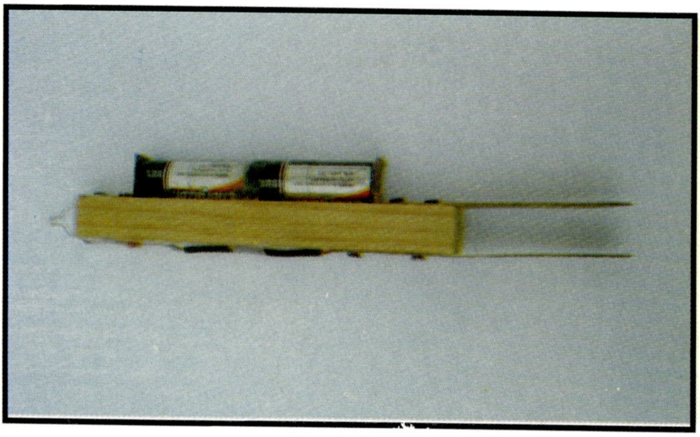

LOW-COST GALVANOMETER

This is made simply from a piece of ply wood, a straw and an induction coil. It is used to measure the magnitude and direction of an electric current.

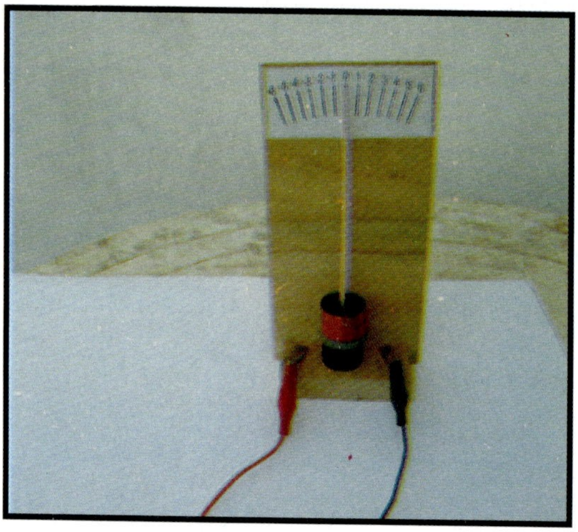

LOW-COST ELECTROSCOPE

This apparatus is made from a plastic bottle with a cap, a pair of aluminium leaves instead of gold leaves. Used for detecting and measuring electricity (especially as an indication of ionization of air by radioactivity)

CLIP MOTOR

The motor is made from simple materials such as a paper clip (from which it derives the name), a wire, a magnet, thumb screws and a piece of wood as a base. It is used to explain the how a motor works.

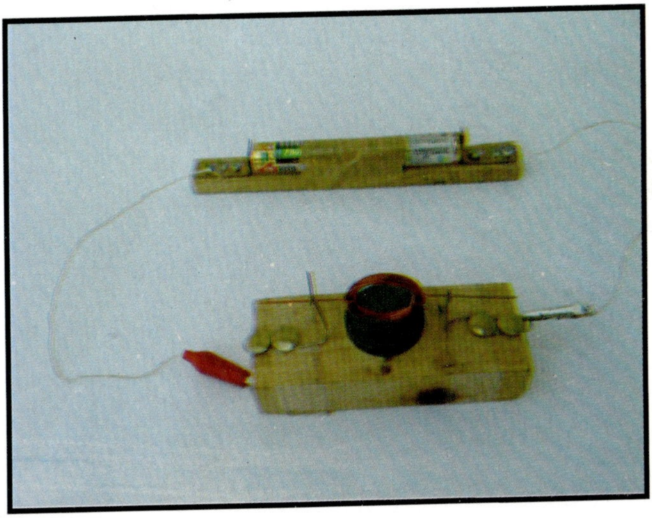

ENGINE MODEL

The engine model is made from chip/particle board. It is used to demonstrate the different stages of strokes in the engine.

ELECTRIC PENDULUM

Made from a steel wire, wood base, cotton thread and two carbon materials at either end of the thread. It is used to study the nature of charged bodies and electrostatic induction.

REFERENCES

Ministry of Education (1996): Educating our future, Zambia Education Publishing House (pp 35-36).

Naidoo R. and Bwalya TM (1995): A Secondary Geography of Zambia, Longman Zambia, (pp1-5).

Philips [The Royal Geographical Society.(2001)]: World Atlas and Gazetteer, George Philip Ltd. (pp 48)

19

Future Prospects of Science Centres and Museums in the Changing Economies and the Ongoing Technological Revolution in the Developing World

Jibananda Mishra

Centre for Science and Technology of the Non-Aligned and Other Developing Countries (NAM S&T Centre),

Core 6A, 2nd Floor, India Habitat Centre, Lodhi Road, New Delhi 110 003, India

Phone: 91-11-24645134, 24644974, Fax: 91-11-24644973

E-mail: mishra_nam@rediffmail.com, m_jiban2003@yahoo.co.in

INTRODUCTION

Science matters to every single one of us. It affects our everyday lives in countless different ways, and the scientific advances of today will determine how our lives change in the future. People around the world are interested in creating wealth, using and adapting technologies for their own needs, having a say in democratic processes, using new media and engaging with popular arts, literature and culture. Particularly for the young, science and modernity are inseparable, and technology is culture. Images of a modern developing world are unthinkable without science, engineering and technology.

THE HISTORICAL DEVELOPMENT OF MUSEUMS AND SCIENCE CENTRES

Museums were a 17th century innovation in the world. They had their origins in the collections of merchants and explorers. The first of these collections to be exhibited to the public in England was at the Ashmolean Museum in Oxford in 1683–a museum that had its origins in science and formed the basis for science as a subject of study at the university. Lecture-demonstrations became a fashionable vehicle for science communication in the 1800s–where these events were very much a part of the cultural climate of society akin to, say, the theatre. During the 1920s and 1930s, the history of science came to acquire new importance in the universities and new popularity among the masses. Many academic histories of science departments in the universities find their origins in the inter-war period, and most of the science museums opened or re-opened their doors to great success during this period. Yet the widespread interest in science history was not uniform. Although these different contexts and their practitioners shared the common ground of positivism and the experience of war, nevertheless their focuses, practices, associations and audiences differed greatly. One area of outstanding differences was attitudes toward material culture. Naturally, practitioners in the science museum took greater interest in material culture than their counterparts in the academy did, but these differences in interest cannot simply be attributed to relative proximity to or distance from scientific instruments. Rather, differing attitudes toward material culture among different groups of practitioners must be understood in the context of contemporary view-points about the role and importance of science history. These view-points involved important questions as to the place of science history in society, the appropriate identity for a practitioner and his/her audience and the role of science history in the post-war work of reconstruction.

Tremendous changes have taken place in science centres and museums in our century. They may be characterized by two features:

- The interaction between visitors and objects exhibited has been growing with ever-increasing pace (dioramas, push-button, working replicas of historical artefacts, exhibits); and
- The conservation of historical instruments and devices is no longer considered a privileged road for the propogation of scientific culture (Science Centres). In the first half of the 20th century this trend has been encouraged by the positivistic and neo-positivistic approach to science, and in the second half of the 20th century by the philosophy of 'learning by doing': the results have been a neglect of the history of science and instrumentation.

During the past decades a debate was started by a few about the inevitability of this trend. Others have begun to show in concrete terms how history of science and museum collections can at the same time be a qualified environment for both the preservation of scientific instruments, as well as for the propagation of scientific culture.

THE ROLE OF SCIENCE CENTRES AND MUSEUMS

Science Centres and Museums as a tool

Museums of science and technology have had an important role to play in the world of science communication, developing 'hands-on' exhibits to convey the principles of science and technology and the social and economic issues surrounding their actual practical applications, alongside undertaking more scholarly and curatorial activities. The great museums of science and natural history have led the way in adding explanatory displays to the presentations of artefacts, and have in essence reinvented themselves over recent years, both in their displays and activities, and through their websites. The science museums describe the Welcome wing, which acts as 'a breathtaking theatre of contemporary science'. It encourages strong visitor participation in the assessment of future technologies.

Science centres have the role of public explanation and engagement without the historical and curatorial roles of the museums. The responsibility is to the fundamental science of particle physics and astronomy, which has few short-term applications in everyday life. This makes its role in science communication particularly challenging. 'Science communication'–an interactive process of public engagement–is the aim rather than the explanatory mode associated with the use of the term 'public understanding of science'.

Therefore, the role played by science centres and museums in particular is of utmost importance. As a matter of fact, the interactive nature of the science centres/ museums represents a unique opportunity to stimulate public interest in science. According to a recent survey of science museums' visitors, it appears that above all they are convinced that the information they get during their visit is correct and reliable. This is something science museums must preserve at all costs; it is not easy to gain public confidence. It is a task that is difficult to fulfill and needs skills, organization and a long-term perspective.

In this context, I would like to underline the need to raise public awareness of science and technology and to improve the communication of science. Nowadays, the relationship between scientific progress, economic development and social change is closer than ever before. Given this dependency it is worrisome that public interest in science is relatively poor, and that it is usually triggered only by health and security issues [hazards of nuclear power, biotechnology, risks of electromagnetic fields, BSE (mad-cow disease) epidemic etc.]. Increasing public awareness of science and technology at all levels, regional, national and international, should therefore be a fundamental part of the implementation of any and every research area.

From Tool to Independent Sector

Too long have science centres/museums operated in the shadow of the other groups' preoccupations: of scientists, of formal educators, of politicians, of technicians and industrialists. The argument is not that we cannot render service to all these disciplines. We will certainly continue to do so, as we have done successfully in the past. The argument is rather that science centres/museums should never become mere derivatives of problems outside their direct influence. We cannot cure scientific illiteracy (whatever that may be) through a family visit of two-and-a-half hours. Nor can we solve the identity crisis within the formal education system, comparing the 1,600 hours a child spends in school each year with perhaps five or so hours spent in a science centre.

But science centres/museums can look back over 200 years of development and experience, as the hosts of large crowds and as media of communication. They have gained considerable expertise in the role of intermediary, interpreter, and as informal learning environments poised between science, technology and industry on the one hand, and the lay public and schools on the other. As the daily flood of information increases, science centres/museums have developed innovative ways of compressing and processing data into wonderful and exciting activities, demonstrations, programmes, theatres and exhibits for a great variety of target-groups. Science centres/museums have gained an identity of their own and are extremely well equipped as public institutions to play an independent role in time to come.

The Science Centre as a Forum for Discussion

The coming decades will be dominated by global issues: the environment, population growth and, perhaps most influentially, the globalisation of the economy. With 12,000 new inhabitants per hour, nature will increasingly come under pressure of culture, and mankind will be forced to pick up the burden of management of the planet earth. At the same time, never-sleeping global cities have been born, amongst which billions in money and data are continually being exchanged. We are continuously being informed about events everywhere, at every moment, in every detail. The security of stability is also gone. Two or three major technological revolutions occur within one lifetime. The accelerating pace of change creates opportunities, but also tensions: in the labour market, amongst different generations in one family, and within individuals trying to cope with the continuing rise in productivity. It pushes many into a passive consumer role, sometimes temporarily, sometimes structurally. These global changes and along with them major shifts in types of employment and class distinctions impact the basic conditions of mankind. Some philosophers point out that man has created a complete, new environment. They distinguish between the production of objects which are immediately consumed, and other objects that last much longer and therefore form a complete new environment into which new individuals continue to be born. The latter includes computers, television sets, houses, cities, the vast man-made changes to our landscape, modern means of transport, and even science centres. These "objects" form a new "techno-tope" in which a majority of the mankind lives, works and dies; and philosophers argue further that human consciousness is being conditioned by these "objects". Or more precisely, they argue, that we are being conditioned by the processes these "objects" engender, in terms of a continuing, partly automated, rise in productivity. We have become, they say, a society of labourers, entirely focused on the rise in productivity, which has become an end in itself.

They also argue that those who create these enduring "objects"–the "toolmakers" of our society–are thus conditioning mankind, despite the fact that the "toolmakers" are only guided by principles of utility and only value "objects" they create in relation to the market price they might command. And further, that "toolmakers" tend to objectify everything: nature, animals, men are tools, and now even the earth itself is merely a launch-pad for the conquest of the universe.

The philosophers criticise this process of objectivisation and state that man has lost all his, individually unique, humanity, which distinguished him from animals. They say that man, whether as worker or as toolmaker, has engaged himself in cyclical, almost "biological processes", just as animals are solely engaged in the never-ceasing fight for reproduction and survival.

I am always inclined to look skeptically at such theories, especially when these are propounded with a certain moral undertone. Nevertheless, I do believe that many of these philosophers very sincerely questioned the role of technology and science in our society, which is very relevant. Our society, endlessly and very hectically engaged in the rise of productivity, lacks public places for contemplating the social impact of technology, science and industry. In the west, churches once played that role in society, being not only places of worship, but also places for public discussion about scientific issues, and market places where the toolmakers of society would meet.

The Contextual Framework for Public Discussions in Science Centres

Science museums and centres have always been children of their time. They reflected the modes of production, the ways in which we, the workforce, were organised and the education system that went along with it. Some also tried to counterbalance the prevailing trends of their time, by setting out their own course and by defining themselves as agents of change within society. If science centres and museums want to engage themselves as agents of change in our present society, they should take active part in the current debates and discussions about the social consequences of scientific, technological and economic developments over the recent decades.

Many publications argue that these developments have produced new socio-economic divisions in society. And new types of jobs and work and, along with them, great differences in income. And in people's perspectives on their lives. At the same time, it is noted that our cities are passing through major crises, as is our formal education system. And many authors state that all these changes and problems are very much inter-related.

The Educational Context

Although our competitiveness and economic prosperity depend on intellectual capital, not everyone can become a symbolic analyst. Nevertheless, the same opportunities should be available to as many people as possible. And so it is argued that educational budgetary provisions should be greatly increased. Special attention should be directed towards the non-material aspects of the economy: lifelong education, and enhancing people's skills in respect of new technologies.

Schools are first of all criticised for not teaching pupils the basic skills of reading, writing properly and calculating. Nor do they turn out knowledgeable pupils: many people today are considered scientifically illiterate because they do not know/

understand basic scientific principles and/or facts. Yet schools are also criticised for hammering away too much at facts and figures (high-volume oriented education), for failing to encourage creativity, failing to foster young people's interest and enthusiasm for science and technology, and failing to develop their intellectual skills (necessary for a high-value economy). Secondly, schools are overwhelmed with all sorts of society-oriented programmes. They are required to touch upon ethical issues, cultivate civic culture, repair the crisis within the family or the neighbourhood, make people more politically aware, stimulate career awareness, etc. Yet, at the same time, many critics accuse schools of having lost their prime position as places for the transfer of knowledge, and the development of insight. Of course, it is argued, schools will never become outdated. Primary schools in particular, will always be needed as places where children learn to master basic skills. But secondary schools are beginning to look "endangered".

Many point out that rising individualism and the enormous increase in personal freedom, have created tensions between the formal culture of school and today's youth sub-cultures. Schools have become less and less relevant to youngsters, faced daily with increasing numbers of choices in an increasingly complex society.

The Information-age Context

Some state that we are entering the silicon age, after the stone, bronze and iron ages. They are referring to the chip–that small piece of silicon on which we are now able to produce several million integrated circuits within a few square millimetres. In the next decade we will be able, through the use of the submicron technology to produce a completely new generation of computers, incomparably more powerful than those of today.

At the dawn of the information revolution, information technology created its own industrial and economic structure. But since computers have penetrated almost every field of work, any new technological development diffuses rapidly. Similarly in the 21st century, biotechnology will revolutionise every field with a biological basis (pharmacology, health, agriculture, food processing, chemistry, environmental technologies and more..). Actually, the fusion between these two "mega" technologies can already be seen (in biometric decoding, for example, to automate passports or replace pin codes), accelerating one another. And perhaps we will have to think seriously about such crazy ideas as bio-computers, built from artificial bio-matter into which liquid software is poured, as images "stream" into our own neural networks through our eyes and senses.

Sub-micron technology will play a major role in stimulating optical and microelectronics: it will greatly increase the amount of information stored on tapes and CDs, as well as the amount of data transmitted through optical fibres (intelligent highways). The new generation of computers is also expected to expand opportunities in computer simulation, by creating "Virtual Environments" (VE). At the moment, most applications are in the area of flight and driving simulations, but the scope will be greatly extended.

Looking for other mega-trends, apart from information technology and biotechnology, that will accelerate our present technological revolution, consider also the developments in the field of micro-mechanics, the use of super-strong fibres (in spite of their present commercial failure) and the introduction of mobile telecommunications on a global scale. Add to these mega-trends the 24-hour real-time global consciousness, mentioned before, and the revolution in our intellectual tools, and one starts to wonder: how can we keep up?

Comparing the amount of new information brought to us daily between, let's say, 1850 and 1900, and between 1950 and 2000, the increase is breath-taking. Rademaker already applies the concept of half-life to the knowledge of someone who has just completed his studies, and this again stresses the absolute need for continual up-dating and lifelong education.

Further, as worldwide mobility continues to increase, we will move more of our urban infrastructure underground to keep our cities moving. Technology is also becoming "invisible", hidden away in privately owned laboratories, or packaged for home use in mysterious softly-humming boxes.

The problem then seems to be threefold: increasing specialisation, the (increasing) amount of information itself, and the invisibility of the processes that lead to our scientific and technological end products.

The Challenge from within the 'Science Centre' Community

Last but not the least, science 'centres are being challenged by their own successes. In the last two decades many new science centres have opened their doors and if one walks through them, one often encounters the same themes, the same type of exhibitory, all over the world. We have still to learn a lot and improve our concept which would inspire us in the next decades to innovate, though, not to replicate as we have done throughout the history of scientific and technological culture. In summary, these rich science-centre concepts are too magnificent to be turned into stereotypes.

The Need of the Hour

The changing nature of the world economies and the ongoing technological revolution require a technologically skilled, and scientifically literate labour force. Scientific skills are a fundamental general requirement in this regard. The developing world has the opportunity to play a leading role in research and development in numerous fields, including medical, aeronautics, telecommunications, computer science and the automotive industry. To do all these, we need a new brand of a learner–skilled in analytical thinking, continually questioning and reasoning current methodologies and always looking for better ways of doing things.

Our vision should emphasize creation of world class science centres and museums for learners and educators and the public, providing edutainment and interactive exhibits, career mentoring and in-reach and out-reach programmes. In this context, the science edutainment centres and museums should focus on the key factors such as experiencial learning, dynamic exhibit policies, future-centric approach, open-minded inclusion of arts and culture, entrepreneurship, quick reaction time to new needs, and the constant re-examination and addressing of public needs. The long term and important benefits of our science and technology strategy should be focused on the role of science centers in promoting this strategy for the social and economic uplift of developing countries as a whole.

To tackle some of these problems, a Commission like body should be set up and within its Framework Programme, the activity called "Raising Public Awareness of Science and Technology" should be included. This activity will be aimed at bridging the gap between science in all its dimensions and the public. This will help the citizens of the developing nations to better understand both the beneficial impact of science and technology on their day-to-day lives, as well as limitations and possible adverse implications of research and technological developments. The activities also aim at increasing scientists' awareness of issues and subjects that are of concern to the public and to promote a better dialogue between science and society.

Creating a science communicators' forum for exchanging and diffusing science centres/museums, best practices would be successful in making the collaborations. The forum will focus on key challenges in the field of economic and societal change, and developments in communication.

Establishment of the Science Learning Network (SLN) in the developing world would be an online community of teachers, students, schools, science museums and other institutions developing resources and activities for science education. This will incorporate inquiry-based teaching approaches, telecomputing, collaboration among geographically dispersed teachers and classrooms, and Internet/World Wide Web content resources.

INTERNATIONAL EXPERIENCE

International experience has clearly attested to the contribution of science centres and museums in enhancing learner's and society's scientific cognitive abilities as well as contributing to urban renewal and tourism. As has already been demonstrated through other initiatives, this exciting initiative can only succeed if government and the private sector work together in partnership. We must call on the private sector to work with us to equip us with the skills that are required to achieve our development objectives. Not only will this provide the private sector with opportunities to develop the competencies of the future workforce, it will also provide creative and dynamic opportunities for the private sector to visibly demonstrate and showcase its commitment to the development of science and technology through active sponsorship of projects and programmes in the science centres and museums.

The private sector, like government, has for some time recognized the importance of this field of contribution. Many companies contribute and promote this field of learning and now we can provide an excitingly new dynamic additional area in which we can know that–working together–we can create a win-win situation for the community, and the developing world as a whole. But more importantly, at an individual level, we help to equip our learners with a real opportunity to make a difference in their lives and the lives of others. We equip them with the skills they need to contribute at meaningful levels and we equip our economy with skills that allow us to compete more effectively in the global knowledge economy.

I am confident that with all the work that we do–focusing on science and technology–the learners who will emerge from schools will have a solid foundation which will enable them to become the future engineers, accountants, entrepreneurs, scientists, inventors, biochemists, geologists, innovators and programmers.

CONCLUSION

I would like to make some concluding remarks. To my understanding, documents should be prepared having an aim to launch a wide-ranging debate on the role and place of science in our society. It should concentrate on four aspects: science-society relations; the definition of research priorities; risk management; and dialogue between scientists, politicians and individuals. By proposing actions to be undertaken or planned at regional, national and international levels, the paper should ultimately help our society to make better use of scientific and technical progress.

This paper will be another step forward in the development of the Research Area, which attaches a special importance to the relations between science and society.

I would like to stress that more actions from our side are required if we want to achieve the objectives I have indicated during this speech.

It is high time for the developing world to seize upon its specific historical and cultural traditions by turning the tables and reconsider the very place of science within culture and the role scientists are asked to play in society.

Brief Report

Science is a part of our culture and society and the recent years are witnessing unprecedented growth in various fields of science and technology, such as materials technology, medical sciences, biotechnology, information and communications technology, etc. In order to fully comprehend these developments and their impact on the socio-economic progress, particularly in an emerging economy, the science centres and science museums play a vital role because visual impression helps in the process of quickly grasping various facets of the new knowledge. Such establishments provide a distinctive networking between the community, the school and the home giving them a unique role in the educational infrastructure. These are often associated with the casual family visit on a weekend afternoon or class field trips, but are increasingly becoming relevant as one of the most powerful tools in science popularization as well as education, such as in the development of major curricula for the schools. Regretfully there are only a handful of countries, which have appreciated the immense effectiveness of the science centers and science museums in educating the young populace and other segments of general masses.

In order to review the status of the science centers, science museums and related establishments in the world, particularly in the developing countries, and to exchange ideas and experiences pertaining to their present and future developments and cooperation, the Centre for Science and Technology of the Non-Aligned and other Developing Countries (NAM S&T Centre) with the approval of its Governing Council organized the 2nd international workshop titled 'The Changing Role of Science Centres and Museums in the Developing Countries' during October 20-22, 2004, in Hanoi, Vietnam, in association with the Centre for Regional Research and Development (CRD) under the Ministry of Science and Technology (MOST) of Vietnam. The NAM

S&T Centre had earlier organised a similar workshop in Kolkata, India, in Nov./Dec. 2002, and a regional Latin American / Caribbean workshop in March 2004 in Bogotá, Colombia. The main purpose of the Hanoi workshop was to bring together science popularization experts and museologists to share the knowledge and experiences in the development of science centres and museums and experiences pertaining to their present and future developments and cooperation.

About 70 Directors and senior experts of science centres and science museums from 15 countries, namely, Algeria, Bangladesh, Egypt, India, Indonesia, Malaysia, Mauritius, Nepal, Pakistan, South Africa, Sri Lanka, Sudan, USA, Vietnam and Zambia attended the 2-day workshop held in Horison Hotel of Hanoi. The nominee of Syria however was not able to travel to Hanoi, but his presentation material was made available at the workshop.

The workshop was spread over six technical sessions and was inaugurated by Dr. Bùi Manh Hâi, Vice-Minister, MOST, Vietnam, Prof. Arun P. Kulshreshtha, Director, NAM S&T Centre and Mr. Phan Huy Chi, Director, CRD, MOST. Twenty-two presentations were made during the workshop respectively by Mr. Bouatrous Ali of the Algerian Embassy in Hanoi; Mr. Mesbahul Alam, Scientific and Technological Advisor in the Ministry of Science and Information & Communication Technology of Bangladesh; Prof. Dr. Salwa Abdalla El-Gharib of the Faculty of Applied Arts, Helwan University in Egypt; Dr. V. B. Kamble, Director, Vigyan Prasar, and Adviser, Department of Science and Technology, Government of India; Mr. Samir Kumar Ray, Director, Central Research and Training Laboratory (NCSM), Kolkata, India; Mr. Ratan Kumar Chakrabarti, Education Officer in Birla Industrial and Technological Museum (NCSM), Kolkata, India; Ir. Herudi Kartowisastro, Director, Indonesian Science Centre in Jakarta, Indonesia; Mr. Viswamitra Oree, Curator, Rajiv Gandhi Science Centre of Mauritius; Prof. Dayananda Bajracharya, Vice-Chancellor, Royal Nepal Academy of Sciences and Technology (RONAST); Dr. Muhammad Rashid Awan, Director, Pakistan Museum of Natural History; Mr. Jayanath Appudurai, Director, National Science Centre, Ministry of Science, Technology and Innovation, Malaysia; Mr. Arumahennadige Piyasiri, Mechanical Engineer in the National Engineering Research and Development Centre, Sri Lanka; Mr. K. M. Lebs Mphahlele, Manager (Science and Youth) in the Department of Science and Technology, South Africa; Prof. Ibrahim Babiker Abdalla, Advisor, Ministry of Science and Technology, Sudan; Prof. Dr. Lé Ngoc Ly from the Naval Postgraduate School in Monterey, California, USA, and Advisor, CRD; Mr. Phan Huy Chi, Director, CRD, MOST, Vietnam; Dr. Nguyên Vo Hung of the National Institute of S&T Policy and Strategic Studies, MOST, Vietnam; Dr. Nguyên Van Huy, Director of the Ethnic Museum of Vietnam; Dr. Nguyên Bá Thái from the National Institute for Education Strategy and Curriculum, Vietnam; Mr. Luu Ngoc Anh from the Educational and Scientific Program Department of Vietnam Television; Mr. Festers Hangandu Mungo, Principal Education Officer in the National Science Centre of Zambia; and Dr. J. Mishra, Programme Associate, NAM S&T Centre. Most of these presentations were the respective country status reports and the perspective plans for constructing new science centres. However some of the presentations focused on other formal and non-formal modes of science popularization, science learning networks, robot contest, science train etc. and Mr. R.K. Chakrabarti of BITM, India, held a fun science show.

The concluding session was devoted to the future cooperation and adoption of a Hanoi Resolution 2004, a copy of which is attached. Prof. Dayananda Bajracharya, Vice-Chancellor, RONAST, Nepal, Mr. F. H. Mungo, Principal Education Officer in NSC, Zambia, and Dr. Nohora E. Hoyos T., Director, Maloka Science Centre in Bogotá, Colombia, were nominated as the respective Focal Points in the Asian, African (including Middle East) and Latin American regions to further develop the activities related to science centers and museums in their areas. National Council of Science Museums (NCSM), India, offered training facility free of cost with individual participants bearing all other expenses and also proposed to offer its Science Exhibition Buses to the member countries of NAM S&T Centre on cost basis; National Council for Science and Technology Communication (NCSTC), India, offered to partly support the participation at the National Children's Science Congress in India, of one / two school children with one teacher as participants-cum-observers from developing countries subject to the approval of government of India; and National Science Centre (NSC) of Malaysia offered training facility in Science Enrichment Programme free of cost with individual participants bearing all other expenses. All the participants urged that more of such workshops should be held in future in this area for all the developing countries once every alternate year, and a regional workshop during the intervening period. In this context, the next regional conference on science centres and science museums was proposed to be held either in South Africa or Sudan or Zambia some time in 2005 subject to the approval of their respective governments and RONAST, Nepal, proposed to host the next international meeting sometime in 2006-2007.

After the concluding session, the participants of the workshop visited the Ethnic Museum in Hanoi and a natural cave situated in Halong Bay, which is surrounded by rich floral and faunal diversity and is an open natural museum that has been recognized by UNESCO as a cultural heritage site.

Hanoi Declaration 2004

While expressing gratitude to Vietnam, the hosts of the international, workshop on 'The Changing Role of Science Centres and Museums in the Developing Countries' jointly held by the Ministry of Science and Technology, Hanoi, Vietnam and the NAM S&T Centre during October 20–22,2004, we, the participants of the workshop resolve that:

1. Science centres shall cover science, technology, engineering and mathematics. These will refer to a permanently established education facility to provide an interactive educational experience through the use of interactive exhibits, displays and programmes.
2. The interactive nature of the science centres and museums represents a unique opportunity to spread scientific awareness, promote non-formal science education, promote scientific literacy and provide life-long learning opportunities for people in science.
3. In accordance with para 48 of the Science Declaration adopted at the World Conference of Science organized by UNESCO/ICSU in Budapest, Hungary, during June 26–July 1, 1999, governments, international organizations and relevant professional institutions enhance or develop programmes for the training of journalists in science, communicators and all those involved in increasing public awareness of science.
4. In pursuance of the Kolkata Declaration of 2002 a regional Latin American/ Caribbean workshop was organized in Colombia, Latin America, in March 2004 and the second international general workshop on science centres for all developing countries has been organized in Hanoi, Vietnam, in October

2004; more science centres in NAM and other developing countries have been established; role of science museums and centres as important elements in public education in science has been promoted; and individual experiences through the quarterly newsletter and website of the NAM S&T Centre have been shared.

5. National authorities and funding institutions shall promote the role of science centres and museums as a vital element in public education in science, and that we shall pursue this with concerned authorities in our respective countries.
6. Science centres in developing countries shall ensure efficient record keeping, conduct regular impact assessments and report progress at regional and general workshops. The NAM S&T Centre will facilitate fund-raising by approaching funding agencies such as UNESCO and UNDP, and other funding agencies and NGOs. There is a need to establish a network among science centres, museums and other institutions engaged in science popularization in the developing countries for effective sharing of individual experiences, and that the quarterly Newsletter of the NAM S&T Centre and its website (www.namstct.org) will report on the developments in the activities of the science centres and museums in the developing countries.
7. Development of science centre and museum related regional programmes be undertaken by the member countries of the NAM S&T Centre grouped in three regions, with Focal Points Prof. Dayananda Bajracharya of Nepal for the Asian region, Mr. Festers Hangandu Mungo of Zambia for the African and Middle East region and Mrs. Dr. Nohora Elizabeth Hoyos of Colombia for the Latin American region.
8. Reviews for assessing impact of science popularization programmes are extremely important for which we may seek the help of the NAM S&T Centre for approaching the international funding agencies like UNESCO, UNDP and also non-governmental organizations/private sector for organizing workshops and undertaking such studies with the support of the NAM S&T Centre.
9. Programmes of Science Centres and Museums in the developing countries may include awareness about and utilization of the traditional knowledge / media for science communication.
10. Member countries of NAM S&T Centre may encourage distance learning network for science communication enabling teachers, students, science centres, museums and other institutions to interact with each other incorporating innovative approaches, internet resources, etc.
11. Exchange of personnel between member countries for providing training in science popularization activities, including the development of exhibits shall be encouraged.

11a. The National Council of Science Museums (NCSM), India, offered training facility free of cost with individual participants bearing all other expenses.

11b. NCSM, India, proposed to offer its Mobile Science Exhibition buses to, or help develop exclusive Mobile Science Exhibition buses for the member countries of the NAM S&T Centre on cost basis.

11c. NCSM, India, also mentioned about its plan of launching a two-year masters degree course on science communication for science museum/centre professionals effective from July 2005. Interested countries may contact NCSM for deputing participants for the course on cost basis.

11d. In view of the importance of India's National Children's Science Congress in encouraging school children to take up investigative science projects, similar activities may be taken up in other developing countries as well. Initially, the nodal agencies of the Government of India like Vigyan Prasar and National Council for Science and Technology Communication (NCSTC) offered to partly support the participation of one/two school children with one teacher as participants-cum-observers from developing countries, subject to approval by the Government of India.

11e. National Science Centre, Malaysia, offered training facility of Science Enrichment Programme free of cost with individual participants bearing all other expenses.

12. In view of the fact that extensive science popularization activities are required in the rural areas of most developing countries, assistance and help be extended to a country where it is needed by the member countries. Awareness programmes/software could also include development of a variety of software on issues like environment, biodiversity, health, medicine, water resources, information technology and so on.

13. More of such workshops shall be held in future on science centres and museums for all developing countries once every alternate year, and a regional workshop during the intervening period.

Zambia/Sudan offered to host the regional conference in 2005 subject to the approval of their respective governments. They will inform the NAM S&T Centre about the venue, date etc. after mutual consultations between themselves and the Centre.

The Royal Nepal Academy of Science and Technology (RONAST) proposed to host a meeting on Science Centres and Museums in the year 2006–2007, subject to the successful establishment of the proper science learning centre in Nepal.

List of Participants

Mr. Mesbahul Alam

Scientific and Technological Adviser (STA), Ministry of Science and Information and Communication Technology, Bangladesh Secretariat, Building No#6, Room No#907, Dhaka, Bangladesh

Phone: +88 02 7167274 (O), +88 02 7287666 (R)

Fax: +88 02 7169606

E-mail: mesbah002@yahoo.com, most@bangla.net

Prof. Dr. Salwa Abdalla El-Gharib

Professor, Faculty of Applied Arts, Helwan University, Orman-Giza, Sarwat St., Egypt

Phone: +2012 3315386, +202 3030433

Fax: +202 3444057

E-mail: prof_salwa@hotmail.com

Mr. Ratan Kumar Chakrabarti,

Education Officer 'C', Birla Industrial and Technological Museum, National Council of Science Museums, 19A, Gurusaday Road, Kolkata-700019, India

Phone: +033 22477241-43(O), +033 24164830 (R)

Fax: +033 22476102

E-mail: bitm@cal2.vsnl.net.in

Mr. Samir Kumar Ray

Director, Central Research & Training Laboratory (CRTL), Block-GN, Sector-V, Bidhan Nagar, Kolkata-700091, India

Phone: +033 23579347-48(O), +033 23573937 (R)

Fax: +033 23576008

E-mail: ncsmin@vsnl.com, samirray52@yahoo.co.in

Dr. Vinay B. Kamble

Scientist 'G' and Director, 'Vigyan Prasar', Department of Science and Technology, Technology Bhawan, New Delhi - 110 016, India

Phone: + 91 11 26864098 (O),+ 9818187061(M)

Fax: + 91 11 26965986

E-mail: vbkamble@alpha.nic.in

Ir. Herudi Kartowisastro

Director, Indonesian Science Center, Jakarta, Indonesia

Fax: + 62 21 3101728 / 3102368

E-mail: herudi_k@cbn.net.id, piaszz@yahoo.com

Mr. Jayanath Appudurai

Director, National Science Centre Malaysia, Ministry of Science, Technology & Innovation, Kuala Lumpur, Malaysia

Phone: +603 20921260 (O), +603 22821197 (R)

Fax: +60320921250

E-mail: jayanath@psn.gov.my

Mr. Vishwamitra Oree

Curator (Electronics/Computer), Rajiv Gandhi Science Centre, Mauritius

Phone: +230 6015238 (O), +230 6273855 (R)

Fax: +230 6862589

E-mail: rgsctf@intnet.mu, mjauabocus@mail.gov.mu

Prof. Dayananda Bajracharya

Vice Chancellor, Royal Nepal Academy of Science and Technology, Satdobato, Lalitpur, Kathmandu, Nepal

Phone: +977 1 5547720 (O), +977 1 4720092 (R)

Fax: +977 1 5547713

E-mail: dbajra@mail.com.np

Dr. Muhammad Rashid Awan

Director, Pakistan Museum of Natural History (PMNH), Garden Avenue Shakarparian, Islamabad, Pakistan

Phone: +051 9219907(O), +051 2280889(R)

Fax: +051 9221864

E-mail: awan48us@yahoo.com

Mr. Arumahennadige Piyasiri

Mechanical Engineer, National Engineering Research Development Centre of Sri Lanka, Ekala, Jael, Sri Lanka

Phone: +094 011 2235384 (O), +094 011 2845653 (R)

Fax: +094 011 2236434

E-mail: nerdc@srilanka.net, techno@nerdc.lk

Mr. K.M. Lebs Mphahlele

Manager, Science and Youth, Department of Science and Technology, South Africa, Post Bag X 894, Room 6108, 188 Schoeman Street, Pretoria, 0001, South Africa

Phone: +27 12 3174391/0 (O), +27 11 8622925

E-mail: lebs.mphahlele@dst.gov.za

Prof. Ibrahim Babiker Abdalla

Advisor, Ministry of Science & Technology, PO Box 2904, Khartoum, Sudan

Phone: +00249 1 83 472360

Mobile: +00249 9 12151748

Fax: +00249 1 83 472362

E-mail: babiker_ibrahim@hotmail.com

Mr. Yousef Khacho

Architect, Director General of Architecture and Museum, Damas Muflurin Shora, Damascus, Syria

Phone: +963 93 709449

Fax: + 00 963 11 2452341 / 2247983

E-mail: ykhashow@ureach.com, antiquities@net.sy

Mr. Festers Hangandu Mungo,

Principal Education Officer, National Science Centre, Private Bag-5, Woodlands, Lusaka, Zambia

Phone: +260 1 263391, 266772 (O), +260 1 250571(R)

Fax: +260 1 263391

E-mail: nsc@zamtel.zm, dsikanyiti@yahoo.co.uk

Mr. Le Van Na

Director, Department of Science and Technology, Son La, 19 To Hieu Street, Son La province, Vietnam

Mr. Dang Van Khan

Director, Department of Science and Technology, Dien Bien, 7/5 Street, Muong Thanh precinct, Dien Bien City, Vietnam

Mr. Nguyen Viet Hung

Director, Geology Museum, 6 Pham Ngu Lao Street, Hanoi, Vietnam

Mr. Nguyen Ngoc Quynh

Department of Science and Technology, Ninh Binh Km2, Ninh Binh, Hanoi Road, Vietnam

Mr. Tran Dinh Su

Director, Department of Science and Technology, Lai Chau, Phong Tho, Cam Duong District, Lai Chau Province, Vietnam

Mr. Dang Canh Linh

Sub Institute of Newspaper and Propagation, 36 Xuan Thuy Street, Quan Hoa Precinct, Cau Giay District, Hanoi, Vietnam

Mr. Hoang Hoa Cuong

Ministry of Education and Training, 49 Dai Co Viet Street, Hanoi, Vietnam

Mr. Le Dung

General Director, PTC Company, Vietnam

Mr. Nguyen Van Lieu

Department of Science and Technology, Bac Giang, 71 Nguyen Van Cu Street, Bac Giang Province, Vietnam

Mr. Nguyen Son

A37, 44 Yet Kieu Street, Hanoi, Vietnam

Mr. Duong Huyen Ngan

Economic Times Newspaper, 96 Hoang Quoc Viet street, Hanoi, Vietnam

Mr. Dao Van Luong

Director, Department of Science and Technology, Ho Chi Minh City, 244 Dien Bien Phu, 3rd District, Ho Chi Minh City, Vietnam

Mr. Nguyen Vo Hung

National Institute Study of Science and Technology Strategy and Policy, 38 Ngo Quyen Street, Hanoi, Vietnam

Mrs. Vu Thanh Huong

Ministry of Science and Technology, 39 Tran Hung Dao Street, Hanoi, Vietnam

Mr. Le Dinh Quang

Centre for Hydrometeorology Science and Technology, Nguyen Chi Thanh Street, Hanoi, Vietnam

Mr. Minh

Director, Department of Science and Technology, Thai Binh, 30/6 Street, Quang Trung precinct, Thai Binh Province, Vietnam

Mr. Chuong Thanh Nhi

Director, Department of Science and Technology, Yen Bai, Dong Tam precinct, Yen Bai City, Yen Bai Province, Vietnam

Mr. Pham Xuan Dong

Institute of Technology Application, 25 le Thanh Tong Street, Hanoi, Vietnam

Mr. Nguyen Duy Sach

Director, Department of Science and Technology, Hai Duong, 207 Nguyen luong Bang Street, Hai Duong City, Vietnam

Mr. Pham Quoc Chau

History Museum, 1 Pham Ngu Lao Street, Hanoi, Vietnam

Mr. Nguyen Ngoc Hong

Director, Department of Science and Technology, Thai Binh, 30/6 Quang Trung Street, Thai Binh Province, Vietnam

Mr. Truong Minh Duc

Director, Department of Science and Technology, Ha Nam, Quang Trung precinct, Phu Ly, Ha Nam Province, Vietnam

Mr. Vu Hao Quang

University of Liberal Art, 336 Nguyen Trai Street, Hanoi, Vietnam

Mr. Nguyen Hong Ha

Director, Department of Science and Technology, Lai Chau, Phong Tho, Tam Duong precinct, Lai Chau Province, Vietnam

Mr. Dao Huy Chung

Director, Department of Science and Technology, Nam Dinh, 1A Tran Te Xuong Street, Vi Hoang Precinct, Nam Dinh City, Vietnam

Mr. Nguyen Van Huy

Museum of Enthnology Vietnam, Nguyen Van Huyen Street, Hanoi, Vietnam

Mr. Ho Hy Liem

Vice Chairman, Vietnam Science and Technology Association, 53 Nguyen Du Street, Hanoi, Vietnam

Mr. Dang Ba Lam

Director, Institute of Strategies and Programmes (Education), Vietnam

Mrs. Tran Thu Huong

Humanity Magazine, 68 Ba Trieu Street, Hanoi, Vietnam

Mr. Dao Tieu Ban

Director, Department of Science and Technology, Lang Son, 438 Ba Trieu Street, Dong Kinh, Lang Son Province, Vietnam

Mr. Phan Hong Tien

Science and Technology Administration Department, Lang Son Province, Vietnam

Mr. Tam Ngoc Ha

Vice Director, High Technology Zone, Hoa Lac, 97 Tran Quoc Toan Street, Hanoi, Vietnam

Mrs. My Huong

Radio and Voice of Vietnam, 58 Quan Su Street, Hanoi, Vietnam

Mr. Luu Ngoc Anh

Education and Scientific Program Department, Vietnam Television, 43 Nguyen Chi Thanh Street, Hanoi, Vietnam

Mr. Nguyen Ba Thai

Ministry of Education and Training, 49 Dai Co Viet Street, Hanoi, Vietnam

Mrs. Thuy Huong

Vietnam News Agency, 5 Ly Thuong Kiet Street, Hanoi, Vietnam

Mr. Pham Thanh Tung

Technology & Service, JSC Company, Vietnam

Mr. Tran Dinh Thieu

Society Science Institute, Vietnam

Mr. Do Van Nam,

Director, Department of Science and Technology, Thua Thien Hue, 18 Hanoi–Hue Road, Vietnam

Mr. Hoang Thanh

Enthnic Development Newspaper, 69 Tran Duy Hung Street, Hanoi, Vietnam

Mr. Thanh Ha

Light Ray Magazine, 51 Ly Thai To Street, Hanoi, Vietnam

Mr. Ha Hong

Peoples Newspaper, 71 Hang Trong Street, Hanoi, Vietnam

Mrs. Nguyen Tuyet Lan

Science and Development Newspaper, 39 Tran Hung Dao Street, Hanoi, Vietnam

Mr. Dang Canh Khanh

Institute for Youth Research, 5 Chua Lang Street, Hanoi, Vietnam

Mrs. Le Thi Minh Ly

Vice Director, Cultural Heritage Department, 51 Ngo Quyen Street, Hanoi, Vietnam

Mr. Dinh Van Ngoc

Director, Department of Science and Technology, Phu Tho, Kim Dong Street, Gia Cam, Viet Tri City, Phu Tho Province, Vietnam

Mr. Dang Van Bao

Director, Cultural Heritage Department, 51 Ngo Quyen Street, Hanoi, Vietnam

Mr. Nguyen Hai Ninh

Ministry of Culture and Information, 51-53 Ngo Quyen Street, Hanoi, Vietnam

Mr. Vu Dinh Trung

A17- Ministry of Public Security, 44 Yet Kieu Street, Hanoi, Vietnam

Mr. Nguyen Mai Hong

News Department–Radio and Voice of Vietnam, 58 Quan Su Street, Hanoi, Vietnam

Mrs. Tran Phuong Hoa

Vietnam News, 4th Floor-11, Tran Hung Dao Street, Hanoi, Vietnam

Mrs. Tran Thu Van

Economic and Urban Newspaper, 21 Huynh Thuc Khang Street, Hanoi, Vietnam

Mr. Thach Can

Department of International Co-operation, Ministry of Science and Technology, 39 Tran Hung Dao Street, Hanoi, Vietnam

Fax: +844-9439987

E-mail: thachcan@hotmail.com

Dr. Thai Van Tan

Senior Officer, Ministry of Science and Technology, 39 Tran Hung Dao Street, Hanoi, Vietnam

Phone: +844-9435376

E-mail: tvtan@most.gov.vn

Mrs. Vu Thi Thanh Huong

Vice Director, CRD & Principal Coordinator of the Workshop, 5th floor, 70, Tran Hung Dao, Hanoi, Vietnam

Phone: 844-9424357

Fax: 844-9421078

E-mail: ttptv@hn.vnn.vn

Ms. Ngo Thi Loan

Project Officer, CRD,5th floor, 70, Tran Hung Dao, Hanoi, Vietnam

Tel: 844-9423875

Fax: 844-9421078

E-mail: ttptv@hn.vnn.vn

Prof. Arun P. Kulshreshtha

Director, NAM S&T Centre, New Delhi, India

Phone: +91-11-24644974/24645134

Fax: +91-11-24644973

E-mail: namstct@vsnl.com, apknam@yahoo.co.in

Dr. J. Mishra

Programme Associate, NAM S&T Centre, New Delhi, India

Phone: +91-11-24644974/24645134

Fax: +91-11-24644973

E-mail: namstct@vsnl.com